Hot Coffee & Beer
For the Project Manager's Soul

Donald A. Pillittere

First Edition

Hot Chili and Cold Beer for The Project Manager's Soul
Donald A. Pillittere

All rights reserved. No part of this book may be reproduced or transmitted in any form by any means, electronic, chemical or mechanical, including photocopying, recording or by any information storage and retrieval system, without permission from the publisher, except for the inclusion of brief quotations in review.

Copyright © 2012 by Donald A. Pillittere

ISBN: 978-0-9851942-0-8
Library of Congress Control Number: 2012912127

Paperback
Published by Donald A. Pillittere
6 Bella Via Lane
Spencerport, NY 14559
USA
dpillit1@rochester.rr.com

Printed and bound in the USA

Cover design and book layout by Leah Balconi

To my family, my wife Lori – the love of my life, sons Donald and David, and daughter Julie, thanks for the joy, laughter, and love you provide every day.

To my Mom and Dad, okay, my father first for the twinkle in his eyes that started my journey and the support of both that led me on such a wonderful path.

HOT CHILI & COLD BEER FOR THE PROJECT MANAGER'S SOUL

Introduction

The general population has many texts that offer help, one of the most popular being the "*Chicken Soup...*" series. But project managers are a different breed, hardened through years of project battles past with the scars to prove it. If you ask a project manager what's the best way to achieve success, most would boldly answer that if a team could be made up of their clones, the world as we know it would be perfect. As much as can be taught through courses, certification programs and various books about project management, real life experience is the second best way to reach project manager nirvana – besides reading this book.

Therefore, there is no way a project manager is going to traverse the aisles of Barnes & Noble or search Amazon for a Chicken Soup for the Project Managers Soul book. No way, no how. Can you image Peyton Manning reading *Quarterbacking for Dummies* as if this is somehow going to help one of the greatest quarterbacks to get better? Or Mario Batali purchasing *Italian Cooking for Dummies*. In these professions, a basic understanding is needed in the beginning, but constantly practicing their craft is what makes them both great.

Then I had an epiphany. I'm blessed with two sons of drinking age with taste for both cold, micro-brewed beer and spicy foods. What would stressed out project managers want at the end of a horrible day of missed milestones, budget overruns, whining team members, irate managers and more challenges on the horizon? - Hot Chili and Cold Beer! After all, no pain, no gain and counteracting a rough day with food and beverage that can do more harm to the body is just what the doctor ordered. Think of it as intestinal training for the upset stomach many project managers experience as they juggle corporate resources to improve revenue and earnings or both all under extreme circumstances.

Project management is not for the faint of heart, but even with all of its problems it's one of man's (or woman's) oldest professions. Even prehistoric brothels needed a pimp. After all a pimp is nothing more than a sharply dressed project manager. Everything in life is a project, getting ready for work, traveling to work, planning the day's meals, kid activities etc. So each and every human being is a project manager to some degree and knows the utter frustrations of juggling day-to-day activities. But the rewards are great, such as driving home after your son or daughter's team won a travel soccer tournament, or seeing them shine in a school concert. Knowing that all of the parental planning you have been doing had a small hand in their success. Yes, being a parent is a life-long project.

All projects come with four key ingredients in my mind, processes, people, parts and phenomena – what I've coined, the Project Management Ps (The PMPs). In order to succeed as a project manager, you must use the people, parts and processes to develop the next great product, while using these same tools to offset and manage the phenomena. Only when these two opposing forces can be brought into balance, can the project move forward to achieve the end goals.

This book includes various writings that provide guidance to the project manager and something to accompany you when nature calls. Many of the stories are from my own experience or ones shared with me by other project managers. Project managers will probably see many familiar situations, and that's good, as we are one big dysfunctional family with similar experiences. Will this book make you a better project manager? Yes, sharing the trials and tribulations of project management is cathartic and certainly can heal the soul. After all that's part of the title so it must be true. In reality, my wish is that there is a reaffirmation of what project manager's already know that works.

I've been privileged to have worked with what two premier project managers. One is a retired Two Star General who mentored me for many years and provided a daily classroom on how to be a project manager. The other person was not only mentored by The General, but continues to impart his great insight into how to be an excellent project manager to me on a daily basis. As a project manager with over 30 years of experience, I can say without question that my appetite for learning how to become better at my craft has never been greater.

Hot Chili & Cold Beer for the Project Manager's Soul

In the Beginning

I'm a firm believer in using simple tools that allow me to perform any job with great efficiency. During my journey in my first book *Are We There Yet, Diary of a Project Manager*, I discovered four common elements of all projects: People, Parts, Processes, and Phenomena. Understanding what is required of a project in these categories, what is available from the team in them and where there are gaps, has been a tremendous way to drive better project success. The project team is only as good as its weakest link, whether it relates to people, parts, processes or phenomena.

So the first write-up in the *Hot Chili and Cold Beer for the Project Manager's Soul* has to be about my favorite tool for managing projects, The 4Ps of Project Management – The PMPs. As you read the book, many of the articles come back to these basic elements, which in my mind can make or break the project, especially dealing with people.

Table of Contents

The 4Ps of Project Management – the PMPs 1
    Process 3
    People 5
    Parts 7
    Phenomena 9
    Conclusion 11

People – The Most Valuable Players (MVP) 13
    Addressing the Human Bottleneck 15
        Can't We Just Fire Tim? 15
        Send Him to The Glue Factory 16
        Workload – "The Current State" 17
        Focus 19
        Rebalancing 20
        Positive Reinforcement 20
        Trade Him 21
        Open The Door 22
        Conclusion 23
    Sense of Urgency – The Missing "Project" Link 24
        Introduction 24
        Where to Start 25
        Leadership 26
        Individuals 27
        Launch Date (The Finish Line) 29
        Motivation 31
        Conclusion 32
    A Few Good Men and Women 34
    Ingredients to a Successful Project Team —
    The 4Fs - Focus, Freedom, Fun and Family 35

How Youth Pushed Kodak into Medical Imaging 35
Ingredients to Success 36
Focus 37
Freedom 39
Fun 40
Family 41
Conclusion 43
The Land of Misfit Toys or the Dream Team 44
Introduction 44
DNA – Dynamic Non-Standard Assets 45
The Extraordinary and Bizarre 46
   Micro-Managing Monica 46
   Kiss-Ass Kate 47
   Analysis-Paralysis Andy 47
   Swiss Army Knife Sam 48
   I'm Not as Good as I Think I am Tom 48
   So Smooth Steve 49
   Political Pete 50
   Rescue the Project Robert 50
   Little Man Syndrome Mike 51
   Quiet But Good Quentin 52
   Phallic Phil 53
   Team Player Extraordinaire Elisabeth 53
   Expert Evan 54
   There has Got to Be a Better Way Gretchen 55
   Arrogant Andy 55
   Talk to Hear My Head Rattle Rita 56
   Not My Job Ned 57

    Figure Head Fred 57
    Workaholic Wendy... 58
    Bombastic Betty or Bob the Bastard 59
    Know-It-All Kirk.. 59
    Not Invented Here Isaac.. 60
    Promise but Never Deliver David.......................... 60
    That's Not How We Do It Howard 61
    Defensive Debbie 62
    Anal Arnie... 62
    Gotta Smoke Sid 63
    Running Ron 63
    Wiseman Wayne 64
    Knob Dick Kevin 64
    Sweet but Clueless Cathy 65
    Underutilized but Content Chris 66
    Cockroach Curt 67
    Legacy Larry.. 67
    Tryant Tammy.. 68
    Oblivious to the World Olsen 69
    Military Mark 69
    Never Respond to an Email Nick 70
    Laid Back Lucy... 70
    Violate the Values Vinnie 71
  Conclusion... 72

Parts.. 75
  The Stealthy Critical Path 76
    In the Beginning 77

    History and MRP 78
    Use the Parts Dealers and Suppliers 79
    Design Practices 81
    Separate Schedule 82
    Take a Chance – Roll the Dice 83
    Conclusion 84

Process 87
    The Process ConundrumGoldilocks and The Three Bears 88
        Introduction 88
        To Process or Not to Process 89
        Phases and Gates 91
        Inventiveness 92
        Trust 94
        Root Cause 95
        Conclusion 97
    Rational Rhythm and Project ManagementHow to Stay in "Project Shape" 98
        Introduction 98
        Roadmap - Product Family Plans 101
        Human Resources – Skills 102
        Institutional Learning 105
        Corporate DNA 106
        Conclusion 107

Phenomena 109
    Can you Manage Murphy? 110
        Introduction 110

Experience – Nothing Like It 112
Time is your friend 114
Camaraderie 115
Don't Open the Door Stupid! 117
Communications or Lack Thereof 118
Conclusion 119

Everything is Interwoven 121
  There's no Time for Eggnog 122
  Introduction 122
  Lead by Example 123
  Family Backing Helps 125
  Intuition 126
  Know Your Children 127
  Trade the Drinkers 128
  Conclusion 129

So What? 131

Resources 132

About the Author 133

Reviews for *Are We There Yet, Diary of a Project Manager* 135

Hot Chili & Cold Beer for the Project Manager's Soul

The 4Ps of Project Management – the PMPs

Every aspect of business comes with the soup du jour of tools most notable by their easy to remember acronyms. The list grows daily as academics and businesses march along creating what they hope will be the antidote that fixes all that ails business. Anyone that has worked on projects has at one time heard or uttered some of the following:

4Ps, 5S, CAPA,

    DMAIC, DOE, FMEA,

        PDCA, QFD, SWOT,

            TIMWOOD, TQM, etc.

Many have become popular because they focus a team in a clear and concise manner eliminating human emotions that sometimes get in the way of developing a strategy or solving a problem. Add to this the many home grown acronyms and teams have the business equivalent of the Swiss Army Knife at their disposal to tackle anything that impedes progress.

What is missing in my opinion is a simple tool that can be used by project leaders to guide them along the project expedition. My personal favorite among the many acronyms is the 4Ps of the marketing mix: Product, Price, Place and Promotion [Because after reading an 800 page marketing text as part of my MBA program, the 4Ps pretty much summed up the entire text!] So during my career when I had the opportunity to go to the dark side – and into a marketing role – the 4Ps came along for the ride. My peers and customers actually thought I was qualified for the job as I frequently opened the 4Ps blade in my knife. Frankly, this simple tool slayed many marketing dragons with great success and focused an otherwise

Attention Deficient Disorder marketer (yours truly).

Given my platonic love for the 4Ps of the marketing mix, I thought it would be fitting to come up with 4Ps for Project Management (The PMPs). And to pay homage to my grandparents Dominic and Grazia who provided another important "P" word in my life – my last name Pillittere. After much thought, especially around words that started with the letter "P," the PMPs were born.

They are:

PROCESSES

PEOPLE

PARTS

PHENOMENA

These are the key elements that must be addressed if you're going to have any chance at a successful project, especially one that you are managing. Project managers are given responsibility along a straight line, with dictator on the left and administrative assistant on the right. Everyone approaches the job with their own personality and within the guidelines provided by the organization and its culture. Companies that provide processes, people, parts and support to address phenomena give the project manager a head start. After this it is up to him or her to lead the team to the end goal – a new product or service that will enhance company profits.

Process

Start at the End

Define Means to the End

Detail the Means

Devils in the Detail

The process is the recipe for that delicious dessert that includes everything required to delight the customer's palate. People are needed not only to make the dessert, but the ingredients, equipment, and other miscellaneous items that are included in the recipe. Parts are the ingredients, mixing bowls, oven, cooking sheets, spatula, etc. Phenomena are the issues that might get in the way of creating a fabulous dessert – cooking at high altitude, the accuracy of the oven temperature, improperly prepared ingredients, using incorrect ingredients and so on.

When you think of something as simple as making a chocolate cake for dessert: with chocolate chips, chocolate covered espresso beans, chocolate frosting and chocolate flakes – there are numerous things that can go wrong. Now replace dessert with a new medical device that is going to be sold worldwide and have to meet hundreds of regulatory standards – and the challenges increase exponentially. The keys to success rely on making time early in the program to address the processes, people, parts and phenomena before jumping into the project. Because it's too costly and time consuming to go back once the momentum of the project has started. Expectations about end dates have already been broadly communicated both internally and externally.

People rarely talk about too much upfront planning. It is always the opposite - if only we thought of that earlier, we wouldn't be in such a precarious situation. It happens when building a house, planning a home improvement project, preparing for a family vacation or shopping for groceries. Why do people pay professionals to build their home or remodel a kitchen? Because professionals know the process, hire experienced people, understand how to procure parts and most importantly how to handle the unexpected phenomena. It doesn't matter how high your IQ is before heading off into project land, if you aren't prepared, you're asking for trouble.

It is learning through experience that makes us better able to manage in the future. Good processes have been fine-tuned over time and encompass those previously missed items that made the difference between success and failure. If you don't have any processes in the beginning and make things up as you go, – the only guaranteed result is a project that is late, over budget and doomed to fail. In today's business climate there is no such thing as unlimited anything, especially time and money to develop processes during a project.

People

Leader

Communications

Old habits Die Hard

Humor Helps

Team members are similar to the opposite sex: you can't live with them and you can't live without them. They're like an extended family where based on statistics, there is going to be an Uncle Raffi – the individual that no one wants to attend or sit next to at the family reunion. However, we do have a choice in courting and then marrying our wives or husbands, something that is not always an option when it comes to the composition of project teams. Project managers relish those programs where the selection of the team is squarely placed in their hands. Project managers rarely get to choose their team members.

A project manager's greatest assets and liabilities are the members of the project team. Success of the project depends on the manager's ability to lead the group, foster communication, mitigate bad habits and inject humor. My definition of people extends beyond the corporation since many projects have tentacles that reach well beyond their own brick and mortar. Sometimes the critical path of the program lies outside the organization; therefore, all people who touch the program are important.

Processes are the great equalizer since they force the many different characters on the team to fall into lock step with the business processes. There is no ambiguity to perform a solo based on your own agenda; each member is expected to abide by the rules as defined by the processes. Look at a football team. Everyone plays their position based on established guidelines and deviation can lead to disastrous results. However, everyone is also expected to help out others when needed. Here a collective group with different skills and roles succeeds if they play within the processes. Processes improve communications, enhance teamwork and foster good habits. The only thing missing is humor.

Humor is a necessary ingredient for projects. Without it, frustrations that come with every project will fester and negatively impact the team's ability to manage the un-imagined. Teams that have fun together also perform better, remove roadblocks and tune into each other and the customers. Humor energizes teams and makes them look forward to going to work each day to tackle tomorrow's challenges. What's that old adage? "Do something you love to do." Working with people that can laugh and joke together goes a long way towards this saying and a successful project.

Parts

Parts is Parts

By now you might be wondering about parts. Am I taking into consideration all of the inputs for success? Do parts include equipment, tooling, test fixtures, real estate, or consumables? Yes, they do. Parts include everything that needs to be planned in order to satisfy the customer. It's just that parts were such an obvious problem with many projects I've managed that it made sense to use (and of course it started with a "P").

The supply chain has become a vital aspect for today's businesses to manage as technological advancements create new products and services in the blink of an eye. Customers demand the best - feature rich products at an affordable price as soon as possible. Market leading companies go beyond this and introduce next generation products in rapid succession. Being able to get a head start on the parts, equipment, and other consumables that are necessary inputs allows the company to reduce time-to-market and reap profits sooner. No longer can teams wait until the product is perfect before revving up the supply chain. Doing so could have parts in transit as competitors are launching the next great product.

Parts are the parallel project that must be managed with the utmost care in order for the corporation to succeed. Adding complexity to the supply chain mix are regulatory certifications like RoHS that can obsolete whole product families. The bottom-line, in today's society, customers don't want to wait for anything. Why frustrate them by taking your eye off the supply chain. Well managed projects land parts just-in-time for production and meet customer demand.

Phenomena

Murphy

Let's be realistic, even if the project was well planned with a tested commercialization process, excellent people, and a parallel blueprint for parts; phenomena or the unexpected are going to happen. So why do project teams continue to fall into the trap of the blue-sky-everything-is-going-to-be-perfect on this project mentality over and over again? First, humans don't want to admit that they're not smart enough to plan for the unexpected. Second, the initial enthusiasm of a new project clouds people's judgment. Third, no one likes change in the form of problems that get in the way of success. What are common human traits carry over to the entire project team.

How do you planned for the phenomena? Well, an experienced project team helps, because they've faced phenomena and have survived to tell their story. Nothing is better than having team members that are project hardened and are not intimidated by common problems. These individuals can educate the team early about potential roadblocks they've encountered in the past and how these issues were addressed. It is up to the project manager to make sure this experience is included in the schedule. Adding time to the project in anticipation of phenomena does no good. Figure out a way to address the phenomena before it can gain traction.

Besides a clear end goal, well-defined processes, and team experience, the last key to mitigating phenomena is attitude. There has to be a certain cockiness or swagger within the team that dares Murphy to cause chaos. The team must take on the Old West Sheriff's mentality that no one is going to mess with my project on my watch. These teams rise to the occasion, quickly harnessing their trouble shooting tools and attacking the problem with the precision of a brain surgeon. How do you create such a team, well, that's the subject of this book? I've been part of both experienced and inexperienced teams that possessed this ability. The only common thread between them was a clear goal and a willingness on the team not to fail in pursuit of the goal.

Finally, sometimes as much as project managers plan for every conceivable aspect, positive phenomena (or dumb luck) can make or break the project. Of course, some people say you are your own luck. These are the times when the whole in terms of team chemistry is better than the sum of its parts!

Conclusion

Are the PMPs: Processes, People, Parts and Phenomena, the perfect antidote that will solve all project issues? Absolutely, after all they are an easy to remember tool that helps simplify the complexities of project management. Eliminating complexity reduces variation, which according to quality guru W. Edwards Deming and others improves quality, reduces expenses and increases productivity. So there's the proof, The PMPs according to 4 out of 5 quality gurus can be used by project managers to reach project nirvana. Just joking of course!

Realistically, project management is always going to come complete with variation: specification changes, team personalities, parts problems, management incompetence, competitive actions and marketing gymnastics. Certainly The PMPS can be used to help minimize potential project issues, but besides understanding how and when to use The PMPs, the project manager also has to assess the quality of these. How good are the processes? Are team skills applicable to project needs? Are suppliers capable of producing parts? Does the team have the ability to take on Murphy?

In the end The PMPs are just another tool that can be added to the project manager's Swiss Army Knife to improve his or her chances for success. They are a checklist of key ingredients that, if properly used, can lead to success. But as with every project, people (both internal and external) create the product specification, develop the processes, make the parts and cause the phenomena. Project success is going to be proportional to how well people involved in the project can be unified as a team for the good of customers – simple as this sounds – it's what keeps project managers up at night.

People

The Most Valuable Players (MVP)

With The 4Ps of Project Management as with all business theories or best practices, people are considered both the biggest asset and the biggest problem. The human being is like no other tool in the arsenal of corporations. After all, the creativity of mind, body and spirit have sent a man to the moon, cured devastating diseases and given society the internet. But even with such immense potential to provide the impossible, there are individuals who are more detrimental than beneficial to a project. Can you say "Brownie, you're doing a heck of a job." In Michael D. Brown's defense, there were many people and government agencies to blame for the response to Hurricane Katrina.

On the flip side there is the human element that rescued 33 Chilean miners. Where the spirit of the miners, technology of many nations, and the support of family turned a disaster into a reason to celebrate. Project managers need to figure out how to take the good variation in people - the ability to adapt to changing conditions, quickly learn new things, and create leading edge technology to improve the profitability of their company. At the same time, they must mitigate the bad variation: searching Craigslist all day for hot deals, taking long liquid lunches, charging overtime for hours not worked, and making costly decisions without all the facts.

People are the single greatest tool at the disposal of any project manager, but the management of such a varied tool is something few have ever fully mastered. Given the impact people play in the success or failure of projects, many of the articles in this book highlight how best to deal with this constituency in order to succeed. However, to say there is a clear, concise, and concrete way to corral people would be lying. The methods and theories to manage people are as diverse as the individuals involved. Let's just say there are guidelines that can help project manager's even deal with Tim – The Human Bottleneck.

Addressing the Human Bottleneck

Can't We Just Fire Tim?

Who's Tim? Well Tim is the adult version of Herbie, the slow moving Boy Scout made famous in Eliyahu Goldratt's book *The Goal* that provides guidance for overcoming system constraints. In other words, Tim is the human constraint or bottleneck that puts the program behind schedule. We've all had experiences with a human constraint like Tim. These are the masters of incompetence that seem to live just under the corporate radar. Everyone on the team knows that their presence is hurting any chances of achieving a critical time-to-market goal.

However, Tim cannot be underestimated for he possesses great powers of prestidigitation and illusion that any magician would admire. Tim is always busy, somehow constantly reminding all within ear shot of the many tasks he is juggling at once. He manages to work on tasks that best suit his interest whether or not they will enhance the project. Tim is a friendly sort and often plays cards or enjoys a rigorous game of volleyball with his peers. He fits in perfectly with any and all non-work related activities. Tim also brings in those fresh hot cinnamon buns on Friday mornings.

Tim is the shiny toy that looks great in the store. He has great features (resume) and speaks well in public. He fools many at first; it is only after months of on-the-job observation that people realize there is not much more to Tim than his bright exterior. In fact, the public persona is what gives the illusion of competence. Tim is hard to pin down and place on a schedule since he is adept at hearing what he wants in conversations and working on low priority tasks. But at performance appraisal time, Tim's plethora of achievements could make a Type A individual cringe with envy. Tim knows how to document anything that is self-promoting.

Additionally, Tim has an unusually high perception of his abilities. Don't get me wrong, skills and knowledge do come with the package, but these are not of the Superman variety that is in the mind, not eye of the beholder. However the skills at Tim's disposal are not always applied to the project in an effective manner. But given an opportunity to showcase these talents, Tim can display them like a Vegas showman.

The problem with our human bottleneck is that the 90-day window for returns has passed and we are stuck with him or her. Or worst yet we've lost the store receipt. This predicament frustrates many smart people because they hired Tim in the first place. Everyone continues to ask the same question, "How did we get fooled? We are smarter than this!" And "How do we remove this human impediment to move the program forward?"

Send Him to The Glue Factory

Just kidding ….. but this is probably the first thought that comes to mind. After all, when an animal is unable to work, putting them out to pasture (or if they're lucky stud) is a routine course of action. However, we are not sending Tim to the stud farm! Certainly in the animal kingdom handling an underperforming mammal is a lot less complicated than dealing with humans. How about that tried and true method of behavior modification made famous in Singapore - caning? No that will get us in trouble with human resources. Almost forgot about them. Maybe they have some guidelines we can use to turn around Tim's performance.

There is something about Tim that brings out the worst in peers. There is really a barbaric reaction that causes even the nicest people to want to quickly remove him. Unfortunately, many times the methods of doing so are harsh. However, the project manager must not fall into this feeding frenzy. He or she must quietly find a way to address the situation. The first problem is that many project managers lack the skill or patience to deal with constraints that involve employees, given all of their other duties.

The best way to tackle this is to integrate the transformation of Tim into everyday activities to make it a natural course of action. You could always take the straightforward approach and call Tim into the office and tell him that his tasks have become the critical path. If he doesn't remedy the situation, he's fired. Nowadays this is just not practical. Human Resource's is still going to require a paper trail to ensure that Tim doesn't hire a lawyer and sue the company.

So where do you start after the initial reaction of sending Tim to the Glue Factory? Well, you have to get into Tim's world - what is he working on? Is there clear direction for his deliverables? Are appropriate timelines set for task? Is he working on the right priorities? The only way to fix a problem is to gather data, yes gather data even with people. You have to understand the current state before moving to a better future state. Without a starting point, both parties will be going nowhere fast. After all, the ultimate responsibility for the success of the program rests on the project manager, even if Tim is on the team.

What follows is a process that helps map out Tim's contribution to the team (or lack thereof). How Tim can positively impact the team, while maintaining his self-esteem. Caution: this process requires data to understand current state, patience to implement the plan, and some gentle cajoling to eliminate Tim as the bottleneck. After that, it's on to the next bottleneck.

Workload – "The Current State"

Maybe we are being unfair to Tim- is it possible that he has too many tasks to complete for one person? Don't laugh, it's a possibility, although far-fetched. In order to find out, a project manager should take the direct approach, sit down with him and go over current and future tasks. Schedule these meetings on a regular basis to understand Tim's role as the bottleneck and to determine what it will take to resolve the problem. It is of utmost importance at these meetings to precisely write down all tasks, start dates, durations, priorities and probabilities of success. There must be a mutual understanding of these items at the meeting. If this is not achieved, the behavior that led to the project's bottleneck will continue. These 1:1s will help determine whether angioplasty or by-pass surgery will be required to remove the blockage.

This contract, and yes it must be a contract, is the start of getting at the root-cause of the problem. Without written data (tasks and timelines), the chances of turning the corner on the project will be greatly diminished and dollars that could be better spent elsewhere will be shredded by delays. It's the starting point for really getting at those capabilities that Tim can provide in order to meet critical timelines. Instead of working hard to get nowhere, the roadmap to the end goal can be defined, with specific roadblocks and trouble spots clearly marked to the path of least resistance.

In *The Goal*, Herbie was loaded down with too much gear and food, so the Scout Leader redistributed his load to improve hiking speed or throughput. This simple move better balanced the process and drastically changed the team dynamics. If we look at Tim's workload, we can shift or redistribute his workload to improve throughput. Easy fixes might not be obvious at the kickoff meeting, so patience is a must. As with all quality tools, the next meeting will provide actual versus estimated progress so the project manager can zoom in on the most critical areas to address. And more importantly better understand Tim's process capability or Cpk.

Focus

After the first couple of documented meetings with Tim, the project manager should now start to get a handle on how well Tim's focus matches the needs of the team. The good thing about writing down tasks, with start dates, durations etc. is that instead of focusing on Tim's shortcomings in a knee-jerk manner, the manager can look at the data. It is amazing to see how people, even if they're part of the problem, jump at the chance to help fix them once presented with data. Data is the common ground that takes out human emotions and allows people to look for ways to permanently resolve an issue.

Given Tim's desire to impress his peers, he could very well be working on tasks that help others in the organization, to the detriment of the team. After all, praise for most teams is minimal throughout the project, typically coming at the end in the form of a luncheon or a hideous bright green sweater. Tim's need for constant stroking which is uncommon for projects, could lead him to inadvertently doing work that provides recognition. Small jobs can add up, and in many circumstances take precious time away from critical tasks. Being able to review the status of each and every task enables the manager to redirect Tim's focus.

Another potential cause for Tim's emergence as the bottleneck is analysis paralysis – overanalyzing tasks with no apparent forward progress. Again, comparing agreed upon tasks from the last meeting, progress to date, and deltas from goals, should bring this to light. The objective of this step is to focus Tim on how he can help the project and the proclivity of the supervisor to take advantage of his skills. The result of this portion of the improvement plan should result in a modified contract that now transitions Tim away from his non-critical to critical tasks. Many business people go to great lengths working on tasks that, at the end of the day do nothing for customer satisfaction or shareholder value. Modifying Tim's contract to focus on profitable team goals is essential.

Rebalancing

The aforementioned patience that this process takes is no more apparent than realizing that Tim's personal capacity for team tasks is not going to eliminate the critical path. Here's where the project leader makes his or her salary, how to reallocate work to other team members or others in organization that Tim can't get done in a timely fashion. Rebalancing brings along other problems that must be managed, such as questions from others who are given some of Tim's workload. Many who believe a better solution would be to escort him out the door, or do what is common for lame horses.

Rebalancing Tim's workload isn't like shuffling chairs on the Titanic. The ship is not sinking because Tim isn't doing equal work. It's really taking more off Tim's plate because his capacity for work is lower than the demand. Without freeing up capacity from someone else, the time-to-market objectives will be nothing more than a dream. In the meantime, the business will consume excess resources and cash and delay any potential revenue.

The project manager can't rebalance Tim's workload and in the process make someone else the bottleneck. To do this right is going to take some time to analyze the team's workload and look for others internal/external that can take up the slack. The prowess to effectively rebalance the various tasks is the characteristic that separates the great project leader from the marginal.

Positive Reinforcement

As with all humans (even Tim), any transition or improvement plan must include a heavy dose of positive reinforcement. It is analogous to potty training a dog, as much as you want to scold them for doing their business on the kitchen floor, you must commend them when they hold off for the great outdoors. So it is with Tim, we must look for ways to praise him instead of rubbing his nose into a mound of delay dung!

The first instinct with Tim is to place your hands around his neck. Don't though. Just remember, Tim brings out the worst in everyone. Positive reinforcement hopefully fosters the behavior we want from Tim. It's the same thing we should be doing in a sincere way with other team members.

It could be in the form of a handwritten note, a small gift certificate to Starbucks (my personal favorite), or a compliment during a team meeting. Simple is better, sincere is better yet; it's not as if we have to hold a ticker tape parade in Tim's honor if he completes a critical task on time.

Trade Him

We have looked at Tim's workload, focused him on important tasks, and rebalanced his load based on his personal Cpk. If our patience hasn't been worn thin by now there is still more that can be done. We could pawn him off to some other department by singing his praises. This would give greater credence to the Peter Principle and fodder for the next Dilbert Cartoon. Or we can truly look at where he might actually find a home and make a positive impact. Some call it human development. I prefer employee Tetris.

Yeah Tetris that popular falling block puzzle game designed by Alexey Pajitnoc that derived its name from the Greek prefix "tetra": meaning four – since all pieces in the game contained four segments. If each employee was a different shape and color piece, how does a supervisor or manager place theses in order to satisfy customers and win the game? To start, one must ask, "What shape and color is Tim and where in the company can he fit? Can his skills be better put to use somewhere else?"

Since there is a soul mate for everyone in the world, there also is a job for everyone, even Tim. It's the project manager's role to help fit the human piece called Tim into the business puzzle. Any employee that has been around a company gains valuable knowledge about products and business processes that can be beneficial within the organization. Therefore, it would be a waste of the company's investment to throw employees like Tim out without looking for a new home. What if there really isn't a place for Tim? Then it's time to show him the door.

Open The Door

Typically the choice of leaving a company is determined by the employee or the employer. The project manager after many attempts at working with Tim to enhance his contributions might decide that the best thing for everyone is to open the door and let him go. If the project manager is lucky, Tim might find his way out before this decision has to be made. I have been in this fortunate position in the past – Tim left before he was going to get fired.

Even if it comes to this step, it certainly can't hurt to re-address the situation one more time – this all depends on the project manager's reserve of patience. Clearly all of Tim's peers would welcome his ultimate departure, however, the project manager's role is different. He or she must make sure that all avenues have been explored before taking the permanent step of firing Tim. Time for one more heart-to-heart talk with Tim to give him one more opportunity to turn things around – this will be the last attempt. Just remember, if you are going to fire Tim, make sure his inability to meet job requirements is well documented, or you might end up facing him in Judge Judy's court! If you want to guarantee this scenario, then drive him to the glue factory – only kidding.

Conclusion

Many project managers have had the pleasure (or pain) of working with a Tim. But can we honestly say that we've worked as hard on this problem as others in business? Unfortunately in business, the people problems are a lot more complex than those related to equipment, parts or logistic mix-ups. However, there is a real business cost of letting someone like Tim go – the knowledge, experience and dollars invested in him by the corporation, not to mention the possibility of a lawsuit.

Next time your blood boils as Tim walks into a meeting and states that he is behind schedule or has spent a week working on an outdated product specification, take a deep breath and figure out what can be done to address the situation. Saving Tim might not be easy, but at least it is worth a try. Everyone has been Tim in the past, maybe not for the same extended period of time, but who can say they've always delivered on time. It could simply be a misunderstanding, a brand new job that is taking more time to acclimate too, or an intrinsic problem with the individual. Following this process can help you determine the root-cause and do your best to address the situation.

Sense of Urgency – The Missing "Project" Link

Introduction

Maybe it's due to my Type A Personality (DNA), but I've noticed something very disturbing about project teams recently; they just don't have a sense of urgency anymore. Not the uncontrolled running from the burning house urgency, but the drive to kick the competitor's ass! In fact over the last couple of years it has become obvious that the sense of urgency amongst my project team members has taken the path of Jurassic Age dinosaurs – a future as compost. It's as if they've lost the battle before the very first team meeting. What was once a vibrant group of people, like fans at an NFL game (Raider fans excluded), has transformed into high school seniors at a library. Individuals who are struggling to complete homework, while waiting for the bell to ring so they can escape and do something more enjoyable.

Has project history with its inherent propensity for delays taken the edge off of teams? Has the frustration of the typical project journey that traveled 25 miles to traverse a mere 10 feet after missteps and redirections pushed team members to reserve energy to survive? Has the constant threat of layoffs and the desire for self-survival, fostered an environment where workers know that hard work doesn't always pay? As US teams crawl at a steady pace, global competitors are racing to the market with superior products.

It has become so bad at times, that team members in close proximity fail to meet face-to-face to discuss issues. Instead they send the emails that end up being unread until something in them is on the critical path. Even the functional barriers that 1980s business books tried to mitigate have come back into fashion. Everyone is waiting for someone, to tell them what to do, when to do it and how to do it. Independence of thought, creativity, and tenacity, has been lobotomized from team members. There is just no energy to speak of when it comes to project teams.

However, ask these individuals how things are going on the project, and what follows is a diatribe about problems. This is quickly followed by a surplus of recommendations for improving the team's success, such as more accurate specifications, more frequent meetings, more ongoing communications, extending help to functions that are behind or bringing in outside support. Ask these individuals why they didn't attempt to implement these common sense suggestions results in nothing more than a blank stare. What causes the person making suggestions to become paralyzed when asked to act on them?

Imagine if Bear Grylls, the outdoor adventurer and star of Man vs. Wild was stranded with these teams. He'd probably end up dying. His only chance for survival would be to babysit them, because most couldn't even pick a banana for sustenance. What evil phenomenon has taken away a team's once proud sense of camaraderie and drive to achieve the impossible? What can be done to transform this listless, ragtag group into a lean, mean, fighting machine?

Where to Start

By now you might be thinking that the easiest path forward is to fire the leader ala George Steinbrenner, hire a new one, spend money to shore up the team and all will be rosy – basically the Yankee's recipe for success. Well if it was that simple, the Yankees would've won another World Series over the last 12 years and failing projects could be easily overhauled. However, we know that project success is analogous to starting a new business or marriage – there is a high probability for a catastrophic ending.

There are plenty of reasons behind team failure and success. The same is true for a team's sense urgency. After all, the combination of unique individuals with different skills, cultures, motivations and personality types is especially suitable for dysfunctional dynamics. So as with most problems in today's society and with the mere mention of George, not Bush or Clooney, but Steinbrenner, a start at the top to undercover a drain of team urgency is in order.

Leadership

Leaders usually get too much credit if things go well and too much blame if things go wrong. In terms of influence on an organization, however, their stamp on a team's sense of urgency is undeniable. Jack Welch preached to his managers that each business unit had to be number 1 or 2 in the market or die trying – his clear message changed the core personality of General Electric (GE). A GE manager with no sense of urgency or drive was immediately sent to the unemployment line. Clearly Welch's influence through senior management and his own vibrato resonated throughout the company. In turn, managers at GE made sure to echo the employee's role in achieving Welch's vision.

Having worked both at a Fortune 500 company and several small ones, my experience has been the opposite of what one would envision. Empowered teams with a drive to succeed were more common place in the Fortune 500 Company than in the small supposedly nimble companies. Each company had a Phases and Gates process to guide the project team, as well as a talented pool of team members. The big difference was the leader at the top of the organization that dictated the strategy and the keys to success.

Maybe I was fortunate to be on several very successful projects during my time at the Fortune 500 Company. Since leaving this company my experience as a project leader has been very frustrating even with skilled project teams. Each time, the leader never really had a clear, well-defined vision and strategy for the company. Product specifications were marginal at best, funding and staffing were inadequate, all decisions had to be made from top-down and the trust in people's ability to perform a job was rare.

Driven teams had a leader that provided a clear objective, adequate funding, freedom to make decisions and trust in the team's ability to succeed even against impossible odds. These ingredients mixed with a decent project leader more frequently led to a timely market introduction of a new product or service. You might ask, why decent project leader and not supreme - because the team, not any individual made the project successful.

So what happens when a project manager is faced with a situation where there is a leadership void? He or she must assume the leadership role and start to work on the ingredients for success – a clear specification, adequate staffing and funding, decision rights and some autonomy. The best way to approach this is based on data. Staying away from the tried and true method of blaming the organization for your problems will all but guarantee failure. The project manager must create realistic dates for key milestones in the project. He or she must define the required resources and anticipate potential catastrophic issues. Too often, project managers don't provide realistic dates because of management pressure. And the result is a late and over budget project. Leadership is about showing the team and management the pathway to success, and a project manager must never try to appease anyone with less than all the facts.

Individuals

Teams that were once dominated by the Type A leader and Type B followers have now been replaced by Type C people. Type C is my term for Comatose: those individuals that attend meetings (mostly to take up space) and only react if led by someone. It is not that these team members are bad or lack talent, it's just that they rarely take a proactive approach to projects or display the resolve needed to compete in today's global economy. While these Type C's continue to live and impede time-to-market, competitors with a greater sense of urgency and survival instincts are pushing them even closer to extinction.

Many times, the project leader is handed a team, based on what the functional organizations are willing to provide. If there is a clear benefit for a particular function, there is a higher probability of management sending one of their A-players. However, even if you are can pick your team, there is no guarantee that success is inevitable. Teams in the business world are far different than teams in sandlot baseball. Picking the best neighborhood kid for your team isn't possible in business. In the business world, there are no Michael Jordans, individuals who can solely make a project successful even if surrounded by lesser player. Basketball has just five players, so the impact of one can be significant. Business teams could be hundreds, making the odds of any one person more important than the rest of the team less significant.

Now Bear Grylls would be a great member of any team. How can you go wrong with someone who entered the Guinness Book of Records as the youngest Briton (twenty three) to climb Mount Everest? Bear is a person that has a calm urgency in perilous situations and always finds a creative way out. What an asset he would be for a project team stranded in the wilderness of the business world.

What can a project manager do with the As, Bs, and Cs? First, work on the weakest links since the team's forward progress is based on their pace. Take time to understand every member's strengths and weaknesses. The goal here is to see if there is something about the individual that could trigger their sense of urgency. Play the role of Dr. Phil and figure out what intrinsically drives each person. It's the project manager's job to match people and tasks in the most appropriate manner to reach the end goal. Good project managers have a key awareness of their players and how to get the most out of them for the benefit of the team. A great example of this is what Phil Jackson was able to do with the divergent egos of Shaq and Kobe in order to win the NBA Championship.

After careful assessment of the end goals and the human resources at your disposal – drastic steps might be necessary. This might be the permanent removal of some team members. Clearly this is not what project manager's savor, but there are times when this is absolutely necessary. The whole team cannot be slowed to a crawl because of one person. No company can afford a team of snails, unless there is some melted butter and garlic around. Never underestimate the impact one negative or incapable person can have on a team. This can sink a team faster than the iceberg that submerged the Titanic.

Launch Date (The Finish Line)

As stated earlier, great leaders not only provide the tools for teams to succeed, but also drive them to the impossible in terms of time-to-market goals. Leaders have to tell teams where the finish line is in the project race. Finish lines that get a collective "Say that again!" or "Say what!" when heard by the team. Steve Jobs at Apple fits this bill because his innovative spirit drives everything at Apple with great success. There is no way Apple could've launched so many innovative products like the iPod/iPad without the employees having a great sense of urgency driven by Jobs. But having leaders like this are atypical in today's world.

Too often, leaders, somehow naively think that what comes out of their mouth for launch dates, even if unrealistic, will happen. I call this the Field of Dreams approach; state it and it will happen. But it isn't that easy. What happens in movies doesn't usually occur in business. Leaders have to understand the effort needed to achieve a particular project and push teams to the limit. Ultimately they need to foster a sense of urgency. However, there are those who are ignorant to the sheer magnitude of a task or short change the team in terms of resources, which defeats urgency from the get-go.

There is nothing more frustrating and demoralizing to a project team's sense of urgency than being handed a "seat of the pants" launch date. Exceptions to this rule are directly correlated to the leader's capability. However, leaders that can inspire a team are outliers – a statistically small portion of the population. Project leaders must still take the launch date and figure out how to achieve these lofty goals, even though the over simplification of projects makes the task quite difficult.

Part of the team's charter should be the ability to intelligently select a launch date based on a specification, appropriate resources and freedom to make decisions. Again, this is a fantasy, but no team ever really receives all that is required to be successful. Still, the project team must take what's handed to them, formulate a plan of attack and commit to a launch date – this has to be a key output of the team. If management pushes back, the project manager should be prepared to go back to his or her team to see if the new date can be met in some creative manner.

What should a project manager do in terms of product or service launch date? The project manager must fight to enable the team to analyze the project and come up with a feasible launch date – one that everyone agrees to. Better yet, multiple launch dates taking into consideration "what if" scenarios in terms of all success, based on past history and with unlimited resources. The ability for a team to own the end date is one of the best motivators I've seen to truly drive the urgency of the team. Teams don't want to be given the ownership of committing to a launch date and then falling flat on a sword in front of management. Teams have a genuine sense of pride because upper management has given them the freedom to utilize their skills to determine what is possible. With this single act, management has leveraged the team's experience, skills and knowledge, and provided an incentive for team urgency.

Motivation

The standard answer for motivating a team and fueling their sense of urgency could be the old standby – money, money, money. However each person's uniqueness in terms of skills and capabilities also comes with uniqueness in what motivates them. One size doesn't fit all, especially when it comes to project teams. Today project managers are also faced with the turtle in the shell syndrome, the turtles (project members) are very scared to peak their head and commit to anything, with the possible implications of being the first to go when layoffs occur. Many find it safer to quietly move in lock step with the herd and hope someone else is eaten by the crocodile.

I've been in organizations that have provided yearly employee bonuses, management by objective bonuses, periodic recognition awards and all sorts of incentives to motivate project teams. Personally, my motivation is taking on a project that has difficult terrain to navigate and succeeding against all odds. The shear challenge of the project motivates me above and beyond any monetary reward. Now wouldn't it be great to have a whole team of my clones? That would make the motivation part pretty easy. As mentioned earlier, picking the players on the team is rarely an option, so assuming there is a way to somehow screen people with identical forms of motivation is impossible.

What can the project manager do in order to motivate a team of diverse characters to drive their sense of urgency and move the project forward? How about taking the KISS approach and asking team members what motivates them. This is a great way to get team interaction on a subject hopefully near and dear to their heart. I've used this with brainstorming rules to enable the team's imagination to run wild. By not squelching creativity, forms of recognition from chartered flights to Paris for shopping and dinner to a round of golf with Tiger Woods at Augusta National were mentioned – with team members trying to outdo each other to roars of laughter!

Putting the fun aside, these sessions always yield some great recognition ideas that can be applied throughout the program to fuel team urgency at various times in the project. In fact some of the teams I've been a part of have had separate recognition schedules that line up with project milestones. Another thing that works is the element of surprise, such as bringing donuts or bagels to team meetings, a hand written note about going above and beyond or an email to a team member's boss about their contributions to the team. My most memorable recognition to date was a note my high school basketball coach wrote about my contribution to the team even though I spent most of the season warming the bench. It was a hand written note sent to my parents that I have to this day.

Conclusion

Question - A highly skilled and motivated team striving to achieve the impossible based on the direction of a visionary leader certainly must have a sense of urgency? Answer - maybe and maybe not. Why? Creating a sense of urgency among teams involves human beings. Unfortunately, humans are flawed (unique) assets that don't always mix into a lean, mean fighting machine. However, if you're on a team, with or without great leadership, engaged peers, excellent incentives and a stretch goal that drives success, you can still make a difference.

Let's call the difference "The Wave," that famous stadium antic where an individual (and some of his or her drunken friends) can get a group of 80,000 to 100,000 fans to do something in unison. Though I'm certainly not condoning drinking on the job or having people perform "The Wave" in a conference room, there is much to be said about the impact one person can have on the team, even imperfect teams that are the norm. It's never easy to be the individual who acts differently from the project herd; in fact if others don't follow it can be downright embarrassing. In the animal kingdom, straying from the herd can place you on the dinner table of a hungry pack of lions. However, as a project leader, your role is to be the one making the bold move to lead the herd.

The project leader has to analyze what is missing from the team and compensate for it so that the team can get back its mojo - a sense of urgency. It can take the form of leading the team, carefully matching team skills to tasks, articulating an impossible but plausible launch date, and providing creative recognition awards along the way. Moreover the project leader must understand that what continues to fuel the fire of urgency will change throughout the program, so what worked earlier, might not later as the team closes in on the launch date. If the project leader is not willing to take a chance by starting "The Wave," he or she shouldn't be in charge! Great players like Michael Jordan want the ball at the end of a game and project leaders must showcase that same confidence and commitment to their team. In a good project leader's mind, no one else is better able to help the team succeed than them. Praying that someone else will take the leadership role is just unacceptable.

A Few Good Men and Women

The previous pages have painted a very bleak picture of project teams, but, this is reality for project managers. People are flawed and it's the ability of the project leader to manage these shortcomings in a profitable way that separates the good from the bad. There are good people and great projects and when they exist, it's one of the few times when the project manager doesn't want to see the project end.

In all honesty, I can say that in over 30 years, there have been just three projects where the team, made me "Jones" for coming to work. These were incredible experiences where a diverse set of people somehow managed to work together. They complemented each other's strengths, overcame weaknesses and took advantage of opportunities to deliver world class products. People worked late, on weekends, and did anything asked of them for the greater good. No one wanted the project to end; we were having too much fun!

What follows is a story about a team that was full of a good men and women, and provided me with one of the greatest experiences of my life. I wish that those that read this book have experienced projects like this or will continue to push forward for the chance to have this rare opportunity.

Ingredients to a Successful Project Team

The 4Fs - Focus, Freedom, Fun and Family

How Youth Pushed Kodak into Medical Imaging

> *"Youth is wasted on the young."*
> – George Bernard Shaw

Well George doesn't know everything about youth - because once upon a time there was a team of young energetic individuals that developed Eastman Kodak Company's first digital medical imaging system.

Ask anyone that has worked in Health Sciences Division (HSD) about their best work experience, and all will mention their time at C Building. C Building was one of many pieces of real estate owned by Eastman Kodak Company and was so named because it had once housed Stromberg-Carlson a telecommunications equipment manufacturing company formed in 1894 as a partnership of Alfred Stromberg and Androv Carlson. Kodak was big on acronyms, and the name C Building was given to this series of interconnected brick buildings.

During the 1980s, Kodak employed well over 65,000 people in Rochester, New York. That put a huge burden on office space. It was during this time that the business unit concept was implemented at Kodak. A group of employees from sales, marketing, R&D, manufacturing, service and financing were relocated to C Building. It was the start of Kodak's entrance into digital diagnostics and those that spent the majority of their time on health imaging products were transferred into HSD.

I was a manufacturing engineer at the time working on products from the 1960s that managed to continue to serve a purpose in medical imaging. Transferring to C Building was scary and exciting at the same time since it was so far removed from the main manufacturing facility. Not long after settling in, I was asked to attend a meeting about a new imaging system – a series of products that would capture images from medical modalities (CTs, MRIs, Ultrasound, etc.) and send them to a laser printer for hardcopy output.

The meeting included my supervisor, a project manager and several department managers. I was handed a letter that spelled out a more senior role – drive product into manufacturing as soon as possible by going "off system" - outside the current processes (Skunk Works). That meant creating bill of materials for boards, procuring parts as they were specified, ordering tooling, and getting products built once design was reasonably stable. Others in the organization from R&D, design, and marketing were hearing a similar tale, an all-out assault on getting the product to market – with attendant freedom to get the job done. Going "off system" was analogous to handing the team the keys to the candy store – freedom to do what had to be done to succeed.

If you were to ask me today if our team was going to be successful, given the huge task presented to us, my answer would have been "Are you kidding me." I am much older now and have seen too many projects fail to meet deadlines. I'm more realistic and cautious nowadays. However, as one of my friends and project team members said recently, "We were young and naïve." In other words, we didn't know any better. We hadn't been jaded by past failures, we had no preconceived notions, we were part of a leading edge project and we took each problem in stride. We were chosen – which meant management had confidence in our abilities to develop their first digital diagnostic system. What a great way to start a project.

Ingredients to Success

I surveyed some of the team members to see if they could articulate the ingredients to our success. Some of the feedback cannot be shared – since our diverse group took great pleasure in afternoon beer luncheons, bowling and everyday ball busting (or whatever the female equivalent happens to be). However, after reading many of the comments, four themes seem to surface. We had a clear *focus* of what needed to be done, we had the *freedom* (within reason) to make decisions, we had *fun* and we became a *family*. Being young and energetic just added fuel to the 4Fs. Our team had the perfect set of ingredients for a successful project.

It helped that we had smart, selfless and skilled individuals as team members. Of course our team was not perfect and there were a couple of Johnson's – you know what I mean. But we worked well with everyone and didn't bother with personalities that got in our way – we worked with them, around them, or told them to their face to get on board. When you are young thoughts tend to come out of your mouth without the brain filtering these comments. However you couldn't argue with what we were doing or how quick it was getting done and our energy and confidence were contagious.

Focus

"Design a system to capture and print laser images from a GE9800 CT"

It pains me to say this about HSD's Marketing, but they understood where the market was going and provided direction for our team. Things weren't perfect and even then our team took great pleasure in giving them grief, but in retrospect, things were not too bad. We had talented individuals analyzing the market to drive products and services that were needed by our key customer - the radiologist.

We had one main diagnostic device to connect to in the beginning: a GE9800 CT. There was no ambiguity with this goal. It is clear to me that part of our success can be traced to a simple, clear end goal to launch a product that could capture images from a GE9800 CT Scanner and print them onto good old Eastman Kodak Company laser printer film. Our team had to push the envelope and break down any barriers to get to market. Units had to produce diagnostic quality images, be reliable, meet quality standards and adhere to all safety and EMI regulations. Well as complicated as the product was – the goal was equally simple. Everyone in management stayed focused on the goal no matter what was happening throughout the project, we had little to no leeway.

In essence our focus on just the GE9800 scanner didn't allow our team to wander or marketing to throw in feature creep. At least this was the case for most of the project. Specifying the hardware/software architecture was less complicated because data on the output of GE9800 was available. The young engineering team even had the foresight to plan out the architecture in the beginning so that individual board design and software modules could be handed over to different individuals. Therefore each and every person managed a critical piece of the overall project – and immediately had a sense of ownership.

Even though the initial modality was GE9800 CT, marketing sold the product to practically anyone willing to pay for laser printer and interfaces. The beauty of the design and initial architecture was that it was flexible enough to handle a majority of these modalities. If we ran into a technical roadblock, our team developed another interface that could do the trick. We didn't let any challenge not even the sale of a non-existent solution get in our way. During this project, our focus contributed to several follow-up product ideas that were successfully developed and launched.

Freedom

"Being empowered to make design decisions without having 47 signatures"

Being away from the main part of Eastman Kodak gave us a sense of freedom, like children home alone while their parents are out for dinner and a movie. To some extent, management cultivated this environment. The team stretched the limits of every process to achieve our goal. Most of the time, it was the only way to keep things rolling. We didn't need multiple signatures to change a design or purchase equipment or parts; we did it as a means to the end. We were given Card Blanche and used this to make critical decisions that enabled us to succeed.

The freedom and flexibility extended to everything in our development process, from product architecture, parts procurement, tooling, assembly flow, factory layout and training. Everyone was just looking for the best way to go about business. Rarely did someone look at the task using their functional glasses; in fact team members helped each other regardless of function. There were exceptions, but we just ran over these speed bumps.

During this time, layers of management were limited, which enhanced our work environment. Many of these managers were also new to this technology as well. Recognizing this, these managers let us go about our business of getting the product to market. At times, the challenge of developing this complicated imaging device was puzzling, but, we continued to learn and move forward. Clearly the managers played a small role in our success, but even without their backing, success was going to be ours. No one wanted to fail because we enjoyed the responsibility management provided for us to make key project decisions.

The ability to make decisions was liberating and rewarding and our confidence grew as the project progressed. This helped as the time-to-market window narrowed and technical challenges had to be quickly addressed. We just didn't think there was any challenge we could not handle and we took them all on! I believe that this freedom to decide was probably one of the keys to our success. We controlled our destiny and didn't have to wait for someone that had no clue about what we were doing to make a decision. Sometimes the concept of freedom conjures up images of a free-for-all. However, if every team member is focused on the goal, freedom, or "empowerment," fosters ingenuity, creativity, issue resolution, resource sharing, and camaraderie. We were *empowered* before the concept even became common to MBA programs and corporations.

Fun

"We worked hard and played hard."

It was a pleasure to wake up in the morning and head to work; even team meetings were something to look forward too. We had a lot of fun, we had a lot of beer, pizza, chicken wings and more beer. Yes, the team that works together and drinks together can succeed. Some team members remain friends more than 25 years after this experience. We didn't talk shop at lunch, just what was going on with family and life outside of work. Talking shop got you nothing but grief and maybe a slap upside your head. Many of us also started families around this time. This brought our team even closer. After all, we worked for Eastman Kodak, the imaging company and could easily share pictures of our children. Not sharing pictures was almost sacrilegious.

Fun extended during office hours between individuals with practical jokes and some brutal humor. Many members earned nicknames that corresponded to some screw up they made during the project or doing an offsite bonding session. No one was immune to this good nature ribbing. New members to the team were quickly initiated into our club – we were an equal opportunity team. The longest nickname for a team member - Homer, Largent, LaLanne, Weber – after Homer Simpson the cartoon character, Steve Largent former Seattle Seahawk wide receiver, Jack LaLanne the Godfather of fitness, and Pete Weber the professional bowler. This engineer took it all in stride.

Major milestones were celebrated, which ended up being just another excuse for team members to kill brain cells. It didn't matter what was provided as part of the celebration, it was just a reason for us to become even closer as a team. We even had Conference Room L – which was our code for the Liberty Restaurant, a favorite meeting spot for various team members. We ate there so often, the waitresses knew what we ordered for breakfast and how we liked our coffee. I think it was awhile before managers caught onto this name as team meetings at Conference Room L became a Friday morning ritual.

Family

"We had two families, one at home and one at work."

We were The Justice League, each person with unique skills and experience with a common purpose. Did we fight amongst ourselves? Of course, but we made up and continued to focus on the task at hand. Even though we each came from different functions our loyalty was to the project and the goal. We hadn't hardened into functional entities that only looked out for the best interest of our career and department. Throughout the project we became a supporting family, looking out for each other and always moving toward our goal. The family atmosphere was another critical piece of our success and the fond memories we had about this experience.

<u>Our team could have easily been called "Characters R Us"</u>

It was the perfect team with each member more unique than the next, starting with Barry. Barry (Brainiac), an engineer who returned from Hawaii to fix a problem that had been plaguing the team for a week – in less than 2 hours. He was one of the young Turks who managed the engineers in parallel with pitching in on the design. Barry could do it all, hardware design, software design or system architecture. As big as Barry's brain was, he was an even more impressive person.

Marty, a former golden gloves boxer and bouncer, was someone who repaired TVs on the side and could fix a flat tire at his house – yes this guy had a tire machine in his garage. Marty could create and fix anything; he was a jack-of-all-trades and an expert at most. He was our version of MacGyver the 80s character that could whip up anything to get out of trouble! His versatility was used by everyone, marketing, design, service and manufacturing. Marty was our go-to guy. He would drop anything he was working on to help us out.

Gary, we called him Doogie Howser (played by Neil Patrick Harris) from the show about a medical prodigy, one of the team's lead production people, someone that to this day is one of my best friends. He kept things going in production and ended up being one of the most requested members by our customers for installing and troubling shooting new modalities. Gary was in such high demand; that I spent most of my time coordinating his many request for assistance from customers, service and marketing. He was the benchmark for customer service.

There were others that made up this diverse but cohesive team that were equally gifted in both skills and character. Each of these individuals handled their functional responsibilities, and more importantly contributed to the team. John, Tom, Randy, and Pete came from design, Joe another young Turk that directed the design effort. Phil, Peter, Sally, Jan, Don and Jim supported production. Toby was our key marketing person -yes we did allow the marketing people in our sandbox and everyone played well together. Lifetime bonds were formed; in fact some team members have been going on a fishing trip for over 25 years. Stories from these fishing trips could make up a novel! The project ended up bringing us together into a family.

Conclusion

Every member of our team continues to have fond memories of their time at C Building, even after 25 years. It was a company family with complementary skills, youthful exuberance, and freedom to decide the means to the end that had a laser focus (no pun intended) on the customer. We were too young to be nervous about difficult challenges. We were ready to prove that management's confidence in our abilities were correct. It amazes me after being on countless projects, some quite successful and others disasters – how those that were successful had the focus, freedom, fun and family cornerstones. Projects that failed were missing one, two, three and sometimes all four of these ingredients.

You cannot go back, but when moving forward you can take what was learned and apply it again and again. All of us, in some way, think back to C Building when we are on a new project and try to bring the lessons forward. They seem to be timeless, give individuals a clear task, have confidence in their ability to get the job done, allow them to have fun and provide a family atmosphere. Chances for success are greatly increased if teams are provided these raw ingredients.

I am thankful for memories of C Building and the friendships, camaraderie and success that came as part of this experience. I've come close to this feeling just once over the last 25 years and hope more opportunities become available in the future. In the meantime, I will have fun, cultivate a family environment, and hope that management provides a clear focus and the freedom for team members to decide their fate.

The Land of Misfit Toys or the Dream Team

Introduction

My father said to me a long time ago, that you should learn something new every day; otherwise, you are not living life to its fullest. Well, my time as project manager has taught me much about the 4Ps (People, Processes, Parts and Phenomena) of project management. My biggest learning so far, even with an enormously talented group of people, if the right-mix of The PMP's is not aligned with the project, epic failure is a strong possibility.

Part of a project manager's job is to somehow organization just the right people on a project to deliver the next product or products in rapid succession to stay ahead of the competition. With more demanding consumers, the ability to employ agile project management while simultaneously getting everyone on the team to act as one is a skill that takes years to master. If in fact this can ever be truly accomplished.

The inherent variability in people that they don't teach in statistics is what leads to problems in managing a project. After all, can you really calculate the mean of Wiley Coyote meeting his due dates, his standard deviation and how to drive him to Six Sigma performance? Well you can, but not many project managers take this extreme step. If this were the case, mapping out a schedule would be more accurate. However, the human animal and their variations change over time and can be impacted by external factors such as health issues, financial problems or non-marital bliss. Even if you thought a previous calculation of Wiley's standard deviation across tasks was valid, these external dynamics will make them invalid.

All projects whether successful or epic failures come with lessons and experiences that can be used on future adventures. The problem is that many people tend to "forget the good, bad and ugly and what got them into those situations" and then repeat history project after project. Even with these unpredictable assets, we call human beings, the good can be better, the bad can be good and the ugly can be beautiful. It just takes time, patience and tender loving care from the project manager.

DNA – Dynamic Non-Standard Assets

Each person is made up of many different characteristics that are rooted in their upbringing, education, experiences, religious beliefs, political affiliations and that damn DNA. We use each of these in varying degrees when dealing with everyday personal and professional situations. In fact, our character is constantly changing, as we age and mature. However, deep down there are core values that keep us grounded. Not all of these core values are good, since statistics tells us that if people's characteristics where placed on a bell curve, the middle section would be normal, but, the tales would encompass the extraordinary and the bizarre.

Given that most businesses employ people aligned with a normal distribution, one wonders how anything is done – since misfits, the statistical outliers do end up on project teams. Of course, the variety of people is endless. Yet, as the project manager you must take these disparate assets and make magic. As with anything in nature, productivity or success is predicated on the weakest link or least common denominator. This forces all project managers to overcome this in order to be successful.

In my experiences as a project manager and team member, I've had the experience of working with both the extremely gifted and the utterly peculiar, the world has to offer. What follows are some of these personalities with names that match their most prevalent and endearing quality. While all people are a mix of these personalities, we all have that one defining character that shines when we part of a team. The project manager's role is simple - connect these dominate characteristics in a way to make money. Sounds easy, doesn't it?

The Extraordinary and Bizarre

Micro-Managing Monica

These individuals do not allow anything to be done without putting their runny red nose into other people's business. Monica has post-it notes about who she called and when splattered all over her office and an endless list of what everybody else is doing. She's to the business world what the producers were to Jim Carrey in the movie Truman. Instead of being concerned about things that might cause a problem, Monica creates problems by bothering people to perform task that are unnecessary or controlling them into a corner. On top of this, Monica talks to people as if they're in kindergarten and she's the teacher.

Monica's need to control others is world class and she tends to need an industrial size bottle of Grey Goose in order to relax. It's hard to figure out why these people have a need to "control" everything to the detriment of progress. Maybe there is a fear of looking bad if somehow a problem is associated with them. By knowing all, the control allows them to filter or modify the message. This is similar to a communist news organization; what is heard makes them shine. However, even though Monica is a pain for those on the receiving end of her micro-managing tenacity (and can't stand her), programs love Monica because she gets the job done.

Some projects need Monica regardless of her personal style. Especially ones that require an eye for detail that is so strict that even one missing data point could be catastrophic. Examples of these are making sure that the raw ingredients in a drug are pure and to the exacting specifications needed to ensure production of patient safe pills or testing of a life support system for Space Shuttle astronauts. Anything that could cause injury or death if not addressed in great detail needs a Monica. Of course as long as her nagging doesn't kill them first.

Kiss-Ass Kate

Lip balm is always on hand for these creatures, as sucking up is Job #1. They agree with the boss on everything and hold onto their opinions until they understand what the boss is thinking. They send notes about how to behave ahead of an audit and cc: the boss to make them look good. They play nice with others to keep the facade, but deep down they're looking for another opportunity to kiss-ass based on the majority thinking of the organization.

What Kate brings to a project is the ability to listen to both sides of an argument, a skill fine tuned over years of agreeing with her boss. Many times, it is very difficult to gather the various opinions and arguments that are part of everyday business conversations, emails, or memos. Kate clearly understands all sides of an argument and filters these into a cohesive understanding of the manager's exact thought process.

Project managers can leverage this ability to listen without emotion and mentor Kate to provide some much needed neutral opinions that could help the team. Sometimes the best asset for a team is someone who can look from the outside in and direct the team towards a decision that will be in the best interest of the customer. With or without the lip balm, Kate can clearly use her active listening talent to help the team.

Analysis-Paralysis Andy

These people, mostly men, always delve into the minutia whenever some problem is brought to their attention. It could be something as simple as the men's bathroom running out of toilet paper that will lead them into Lean Six Sigma mode. Andy then proceeds to value stream mapping the current toilet paper stocking procedure to understand the root-cause of leaving someone's bottom without proper post nature cleansing. These characters want to get way down into the process to discover the Holy Grail of root causes.

The tenacious nature of Andy can be used to the benefit of a project especially when there is a technical issue that has everyone scratching their head or guessing its cause(s). Andy can lead the charge and use his ability to dig deep into the problem to try to uncover the true problem at hand. Instead of guessing incorrectly without a process and causing delays and costly overruns later, Andy can work on solving the problem correctly the first time. It might make for some initial frustrations in the beginning as Andy is Andy, but it will pay off in the end.

Swiss Army Knife Sam

Project managers love these people, because they can always get you out of any jam. Think of a modern day James Bond. When there's an impossible problem- they're called in to fix what all the king's horses and men could only dream of. These people could've even put Humpty Dumpy back together again. Sam just loves the idea of solving problems and doesn't need any ego stroking. It's too bad he's such a rare breed.

Sam, unlike Andy, has the ability to fix things in a hurry without the need to delve into the detail. Whether it's fixing software timing issue or figuring out a wiring problem in a system, Sam allows the team to get back on track fast. He's the project handy man. Project managers need to know whether Andy or Sam is the right person to analyze the problem as they have similar skills but with very different speeds of execution.

I'm Not as Good as I Think I am Tom

All people think they are good at something, but these characters have a distorted viewpoint of their capabilities. Tom tends to be the only one in the company that thinks he's that good (even his wife and children know he's only marginal at best). Tom goes through each day believing that the company would collapse if he were no longer employed. Another problem with Tom, the caliber of his work is often less than required to perform the task.

Unfortunately, Tom does sit on project teams, and takes up space and continues in his belief of self-greatness. All people, even Tom have skills that are beneficial for projects, if the patience exists to find them a place at the table. With global competition and more demanding customer's the time needed to figure out how Tom can best help is better spent elsewhere. Many times the effort is not worth the benefit due to more pressing issues. Therefore, for many teams, Tom needs to be sent to the farm system or sent packing.

Project managers cannot afford to hold the hand of an employee during a fast-paced project to either get them up to speed or find tasks that match their skills. The days of carrying weak links throughout a project are gone forever. No one is going to win a race without bringing the best athletes to the meet and this goes for businesses as well.

So Smooth Steve

These people are level headed no matter what the situation, and manage to get the job done as others around them falter. Steve is very knowledgeable about his job (as well as others) and provides no worries for management. He's as smooth in his apparel as he is in presenting material; the whole persona is just pure silk. Steve could convince Al Gore that there's no such thing as global warming, but chooses not to.

The smoothness comes without any political bent; these characters are just nice to the core. Their purpose is to make the company and people around them better. Steve's place on the project is to articulate the vision as a trusted person that the team respects and admires. If there is one person that could sell the concept, provide the team mantra in a compelling fashion, Steve is your man. With team members aware of the fact that Steve has no hidden agenda and really only wants what's best for the company, the team can move forward with a strong purpose.

Political Pete

Expensively dressed, with star presence and the ability to give great presentation, these animals roam the business jungle. Unlike Steve, Pete is filled with big ideas, such as solving world hunger but doesn't typically follow-up or fix anything. Nevertheless, there is nothing better than to sit in an audience and watch them present with the poise and grace of a master theologian. Interpersonal skills on steroids allow these characters to know everyone by first name, as well as those people's spouses, children, hobbies, and favorite ball teams. Pete is always on and never lets an opportunity to schmooze slip by without taking full advantage.

If there is a problem that needs someone to calm the restless masses or more importantly deal with a customer's frustration, Pete is the man. He can turn a frown upside down or The Hulk back into a human being. The key role for Pete is to speak about how others are going to solve the problem(s). Never let Pete go it alone and guarantee fixes, because this can only lead to other disasters. Someone else has to make sure his campaign promises can be implemented.

Rescue the Project Robert

These characters tend to play the role of Romeo and rescue the damsel in distress (project) from itself, even when some of the problems are caused by them. Whenever there is a team meeting, Robert gives the boss the perfect view of the world thanks to his involvement on the team. He then provides the solution to a problem as the boss listens with attentive ears. Robert tends to do little to help the project on a daily basis and waits to ride in on the white horse or the red fire truck to save the day.

Roberts are typically bright people and do mean well, it is just the timing of their efforts that are questionable. What's amazing is that the boss somehow gets mesmerized by Robert and doesn't connect him to the problem. Unlike Pete, Robert's intellect just needs to be put to use on a more frequent basis. Roberts tend to make great technical leaders, because they enjoy hovering around the neophyte designers and lending their knowledge at a critical time in the program. Too bad Robert didn't offer help in the beginning, but this would take away a rescue opportunity.

Yes, I've dealt with these ...... Characters

As a project manager, you've probably had the experience of working or directing many of the aforementioned characters. At least, you certainly had similar or even better approaches to utilizing them in a manner to support the efforts of your teams. It's the price we pay as project managers that all too often, our teams are chosen without much input from us. Yet in the end, the project manager must mix this group of ingredients into a gourmet meal that will not only taste great, but also fill the pockets of the shareholders and upper management.

If being a project manager was easy, many people would take this on as a career, however, our numbers are limited and our pain is immense. Yet, project managers believe that they can turn the proverbial lemon into lemonade even with some very sour team members! So let's continue with the list of characters and see if this is déjà vu over again.

Little Man Syndrome Mike

The little man with a chip on his shoulder is talked about often and doesn't apply to all vertically challenged men, but it does exist. These people spend the day talking a big game. If in charge, they tend to showcase their knowledge hoping that it will make them taller in the eyes of peers. Mike has an opinion on everything and is not shy about sharing it. Most of his time is spent on trying to compensate for their lack of height to gain some advantage over others. It could come in the form of putting down an idea of a co-worker, correcting someone's mistake, or showing his knowledge of a subject. Mike will be taller even if it is not a physical attribute by making others smaller.

Unfortunately, Mike is a very intelligent person but this gets lost by the constant need to be the center of attention or whine about how dumb everything around him is and what the company could become if he was the CEO. You cannot use Mike as much as you want because once you get through the surplus of ideas, concerns, or intellectual diatribes; there are some great and beneficial pieces of business gold. The only question is how much patience the department has to mind these rare but valuable nuggets.

Mike is a perfect fit for strategic planning or something forward thinking because he'll either know a lot about the subject or will investigate until he is the expert. In addition, given his less than normal verticalness, Mike will make sure there are no stones unturned. Mike will also work his butt off to show what he can do when put into an environment where his immense talents can shine. Therefore, to those that know or manage a Mike, just be patient and you will reap immense riches.

Quiet But Good Quentin

Quentin comes and goes in the day without anyone noticing, but he always gets the job done. He is shy by nature, spending little time chatting with peers. Quentin comes to work at the same time every day and puts his head down and works. He is very meticulous and always completes his tasks on time. A lot can be learned from Quentin as he is well schooled, knows his job inside and out, and can be trusted to be a consistent asset to the company.

Good project managers provide Quentin opportunities to help train others so that the overall performance of the department can rise to the next level. Given his shy nature, the best approach is to convince him that moving outside his comfort zone is a job requirement, since he takes great pride in his job. Sometimes this little push or sleight of hand is all Quentin needs to become an even more critical asset to the team. The world could use more of these talented and hardworking individuals whose only drawback is shyness.

PHALLIC PHIL

There are and always will be, people that are part dictator, part devil and part male organ (maybe his last name is Johnson). They tend to think that bullying others or striking fear in people will get the job done. Problems of their own doing can be easily justified as others and their illusion of being perfect is only exceeded by their poor treatment of people. Many times these individuals continue to survive in business, even as companies promote a caring team environment that values diversity. Guess that corporate America includes phallic as a group in the diversity movement.

Phil can be used once in a while to push people to action, but like your annual trip to the dentist for a cleaning, too much will hurt team dynamics. In a perfect world, Phil is better off in the military where the self of each person is put aside for the protection of freedom, and he will answer to others just like himself. However, in small and strategic doses and with the right owner controlling his leash, Phil can bark and scare the team into action. Too much of Phil can cause nothing but bite marks on team members and tetanus shots to boot.

Team Player Extraordinaire Elisabeth

These are the individuals you want to hug and have an abundance of on your team. There's never a personal agenda and Elisabeth will do anything to help the team succeed. Elisabeth is loyal to a fault and make's everyone around her better. There is no such thing as "not my job" in her vocabulary, anything the team needs, Elisabeth will gladly perform. Even if you asked Elisabeth to paint your house without explaining the reason, consider it done.

Project managers can never have enough Elisabeths, as their commitment to the team is second to none and they have no problem tackling challenges as long as it's for the greater good. The problem is that there are just not enough Elisabeths in the world. In over thirty years in industry, I can list only a handful of true Elisabeth's. Too bad cloning is illegal because project managers would be delighted to have a team of Elisabeths.

Expert Evan

Being an expert and proud of what you know is great, as long as not everything in the world is brought back to this expertise. These people think the solution to a problem can be pigeon holed into their expertise when in fact the opposite is true. Evan, if left unattended, will cause the project to come to a screeching halt, as he take hours, weeks or even months to solve a problem with the wrong tools. Even when the problem lines up with his expertise, time is lost as he tells everyone how much he knows about the subject without ever applying it to the actual problem.

Six Degrees of Kevin Bacon is a trivia game based on the concept of the small world phenomenon and rests on the assumption that any individual can be linked through his or her film roles to actor Kevin Bacon within six steps. Evan plays this game all the time, as only six steps are needed to fit the problem to his expertise. Does he possess skills, yes, however, project managers have to be very careful at engaging Evan or the application of his expertise will cost the program more money than it's worth.

There has Got to Be a Better Way Gretchen

These characters are always on the lookout for making things more efficient. Some can nicely manage to turn this on when required; others have the button on all the time. Those with control, have the ability to find areas of wasted time and effort. Then get buy in from managers and peers prior to implementing a solution. The other version tends to tell all about how poorly things are and complain, but never take the time to fix the issue.

Project managers need to figure out Gretchen's on/off button so she can be started when needed to improve a process. Without knowing the location of her switch, the effort Gretchen places into her day-to-day work could slow progress as she goes into full improvement mode. A great place for Gretchen is as a member of the corporation's Lean Six Sigma organization.

Arrogant Andy

These individuals are embedded throughout the company and unlike the expert; their arrogance extends to everything they do, both within and outside their functional area. Andy tends to look down on others and to showcase his knowledge even when not needed. Many times his arrogance may have come from some great accomplishment in the past and in constantly reliving this moment. Andy tends to think his dung doesn't stink. Maybe his parents injected him with too much self-esteem and this has carried into adulthood

Well as much as I believe that every single unique person can be used to the benefit of the project manager, I'm just not convinced that Andy really belongs. If the company promotes funding entrepreneurial efforts, the best thing to do is play to Andy's ego and see if he can truly live up to his lofty opinion of himself. Otherwise, stay away from this draft choice when it comes to picking the project team.

Talk to Hear My Head Rattle Rita

These individuals don't need an iPod; just the sound of their voice brings them comfort. Maybe the vibration equates to a sexual experience. The problem with these people is that they waste time; there is never a simple Yes or No answer. Rita is well meaning and doesn't have a mean bone in her body, but she can talk the ears off a tin horse.

When Rita finally does get to the point, her contributions more times than not do add value. Nevertheless, the time taken away from the team to perform other tasks as Rita pontificates can lead to frustration. What Rita needs is time to wrestle with all of her ideas prior to a meeting so her talk time can be limited. Never, ever, ask Rita to think on her feet, unless you have plenty of military rations to survive.

I just don't have patience for Rita on my team. Even though she can provide value, trying to minimize her need to talk just takes excessive time. Rita would be better off in a think tank, where people are paid to exchange ideas verbally.

Enough I can't take it anymore!

Go for beers with a group of project managers and the majority of the conversation will center on people – it's always been their greatest challenge. Even if a project manager could pick his or her team, there are no guarantees that the whole will be better than the sum of the parts. After all how can a project manager mine what is the greatest part of a person –

the ability to learn, adapt and advance society without getting some of the DNA that represents the drunken football fan that moons children.

But to the few and the proud, that dare wade into the waters of project management come the opportunities to build a weather satellite, develop a prosthetic hand or come up with a cure for cancer. The best thing that can be done is to look for the best in people, line up task to what they do best and hope that this match leads to greatness.

Not My Job Ned

These characters will not and I mean will not do anything outside their job description; some won't even go to the bathroom if not required. It's not to say that they don't have talent, it's just that it will never be applied outside their narrowly defined job description. Don't ask Ned to work more than 8 hours, because again, not his job to work more than 8 hours unless approved by management with paid overtime.

This one is simple for me as a project manager, no Ned's! Having worked with Ned, his cousins, aunts and uncles for years, there's just no reason to have them on your team. With today's globally competitive world, Ned and his relatives will do nothing more than flush your company's ability to compete down the toilet. Project teams need members who not only do their own job, but are more than willing to do whatever it takes to win – yes business is a competitive team sport.

Figure Head Fred

These individuals are the business equivalent of the Hummel figurines; they look good, but don't serve a purpose. Even in today's business world with less layers of management than previous norms, these characters still exist. Fred does not possess the knowledge required to do the job of his subordinates let alone to manage them. Somehow, out of the blue Fred lands a management position to the chagrin of everyone in the department.

Fred is very skilled at moving onto his next job without actually doing anything of value on his previous assignment. He is a vertical climber that dodges anything that could negatively impact his image or resume. Even though the majority of people in the company clearly see through Fred, he's a business ninja. He blinds those in charge of hiring him.

How does Fred fit into the project? Many times he is made the leader and typically above someone who is actually skilled at project management. The problem, Fred overrides the capable project manager and puts the team on the fast track to pull a "Thelma and Louise" (if you haven't seen the movie, it's worth a watch). The subordinate project manager has to carefully go around Fred and quietly undo the damage, while waiting patiently for Fred to get the next promotion. If the subordinate project manager doesn't take on this challenge, the resultant failure of the project will stick to him forever.

Workaholic Wendy

These characters get to work early, leave late and log in after hours. It's hard to tell whether they are actually producing more output for their long hours or are slower and less efficient than their peers. Everyone knows these creatures live at work and management tends to be oblivious to the value all the long work hours provide. Wendy comes in two forms, one that truly does put in profitable extra hours for the company and another that is just doing it for show.

The problem is that management doesn't always understand which version Wendy is and tends to place her in the profitable category. This makes it difficult for the project manager to go against common lore and expose Wendy. For the few workaholic Wendys in the world, any project manager would be giddy to have them on their team. These people are the old schoolers that sacrifice personal life for company and can be nothing but a valuable asset for the team. However, the opposite is true of the evil version, as they tend to shred program dollars at an alarming rate and put the project manager in a bad place. My advice: keep the good, and get rid of the evil, but don't mix them up.

Bombastic Betty or Bob the Bastard

These are the people that no one wants to work or deal with a daily basis; they're just not the friendliest sort. Many times they lay low, but when something sets them off, all hell breaks loose and the person that sparks the fire is usually the victim of 3rd degree burns. No amount of talking to Bob or Betty's supervisor will change their behavior or get them fired, because outside their nasty nature, they tend to do a decent job. Or their boss doesn't want to put the effort into the arduous process to fire them.

If it were my company, Bob and or Betty would be gone in a heartbeat. As the project manager and with Bob's or Betty's abilities to perform well when sedated, it's hard to tread the fine line between the good and bad. The best thing a project manager can do is figure out what makes Bob and Betty tick and see if there is a different part of the organization that can use their talents. Otherwise, the project manager has to protect his or her team from the sporadic tantrums and do their best to promote teamwork.

One last point, in the past, I've been able to find the antidote, someone that Bob or Betty likes and for whatever reason neutralizes or sweetens them. If there is any way to find and then to fit these people onto your team do so post-haste.

Know-It-All Kirk

These individuals are to business what Wikipedia is to the college students, a portal of knowledge. These characters come in two flavors, ones who really know a lot and others whose insecurity makes them want to appear as geniuses to their peers. Think of the know-it-all Cliff Clavin of Cheers and you get the picture of the lesser version of Kirk.

The other Kirk is that out of this world person who never ceases to amaze people with his or her intellect. He's the first person that comes to mind when there is a difficult problem to solve. Kirk never disappoints either; he is just very good at looking at a problem and then going into the deep recesses of his mind to formulate a solution. However, the other version of Kirk can talk a good line, but when the rubber meets the road; just assume he's in neutral. Therefore, the choice is easy for the project manager; pick the version of Kirk in drive.

Not Invented Here Isaac

These characters have been with a company since Jesus roamed the earth and have no other view of business world besides what has been done to date. When outsiders or new hires come with new ideas, these individuals go on and on about how they were on the original team that put the processes in place and how they've been successfully used for years. Isaac's ability to generate new ideas has been depleted or discharged over the years, sucked away by what has always been done.

It's certainly not bad to continue to use what has proven to be successful in the past, however, the business world never rest. Eventually the competition will figure out a better way and put you in the unemployment line. As a project manager, the extended experience of Isaac is always a great thing to leverage since he knows what's gone wrong in the past. The challenge is to take this experience and use it to move the team forward.

Promise but Never Deliver David

These people promise the world, making bold statements about how they will call you, or stop by to address an action item, only to disappoint. Think Santa Claus that never brings the presents on Christmas Day or parents that forget their children's birthdays. These characters are some of the most frustrating people around because even with repeated request for their assistance, they somehow find a reason not to deliver. Yet, when David needs something, it has to be done now!

We have all played the role of David sometime in our lives, if you can remember a late night of drinking and promising something to someone, only to forget about it the next day. David is on this drinking binge at work, not in the literal sense, but by how he acts. Project managers can do without David, there's no need to carry someone who is assigned task, promises to deliver with confidence and then never comes through with the goods.

That's Not How We Do It Howard

You know Howard, he was that teacher you had in second grade that took great pleasure in correcting your mistakes, or the college freshman writing professor that tortured you and sent you to the grammatical underworld. When dealing with Howard, time is wasted prior to the meeting, trying to figure out what might be corrected or changed so you can do so in advance. However, Howard always finds something during interactions to correct or change that is just not warranted. Many times these changes have more to do with how Howard does things than whether it is required to improve the outcome.

There is no problem with doing things correctly, as long as it adds value to the company and shareholders; however, taking this to an extreme is unwarranted. Shareholders don't want to pay for nitpicking just for nitpicking sake. Project managers must use Howard, and his "that's not how we do it attitude" in a way that helps make the company money. If this ends up consuming too much time, Howard has to go.

Having fun yet?

People are as endless in variety as the problems project managers face on a day-to-day basis. As this list of project citizens continues, common themes should become apparent. The project manager's job is to figure out how to turn the land of misfit toys into the dream team, it's just that simple. No one, especially upper management is going to give the project manager a shoulder to cry on because of dysfunctional team dynamics. So the next time you want to talk to someone about your problems, just envision the listener being the Marine Sergeant in the Geico commercials.

Defensive Debbie

This type cannot tolerate any kind of feedback that dare corrects something they are doing, even if what they are doing could cause business Armageddon. They manage to get away with performing their job with little management intervention to make their work output better. No matter what avenue you attempt to help them improve or correct a grievous error, the defensive shields come up and the ears, eyes and brain are closed immediately. Debbie is master at reflecting criticism and holding a grudge for eternity.

Project managers should at all cost avoid having Debbie as a team member. However, if you are ever in this position, your job is to document her lack of performance and cut away this growth from your team. I always believe that there is a way to support and mentor anyone, but sometimes it's better to cut-bait.

Anal Arnie

The perfect office with papers in the exact same position, all pens in one slot (in ROY G BIV order), all pencils in another, paper clips separated by size, walls hangings perfectly level and files for everything including bathroom times and dates for the past 10 years. Arnie is a creature of habit, consistent, persistent habits that drive others to drink. Being anal within one's own world is fine, but trying to force others into this world causes nothing but grief and waste. It's one thing to behave this way at home with your clothes sorted by color, season, mood, brand, and fabric, but a completely different thing at work.

Good project managers figure out what requires this level of organization on the project and engages Arnie to do what he does best. I've seen this done at a medical company, where Arnie was used to collect data for a FDA submission. Let's just say that the package provided to the FDA was a thing of beauty and went through the complicated government process without a hitch. There was no one else that could have pulled this off as well Arnie and his effort led to a better than expected time-to-market.

Gotta Smoke Sid

Even with all of the health warnings, there are individuals that spend an enormous amount of time in the butt hut. It doesn't matter what season of the year- rain, snow, or a brutally hot day- these characters like their ten minute fix. Many are creatures of habit, so meeting with them is not too difficult as long as the time doesn't conflict with one of their many butt breaks. Problem is that it's difficult to get a full day's work from Sid, as his habit in many cases consumes a majority his workday.

If you don't mind the air pollution, much can be done if you venture outside during the summer to talk with Sid – it seems that smoking and thinking go together. In reality Sid is usually not a bad person and can help, but don't expect an eight-hour day.

Running Ron

The former high school athlete or addicted runner is sometimes worse than the smoker. Get out of their way when lunchtime approaches, as plans have been made for a hilly five miler. Obsessed with running or biking, meetings with these folks have to take into consideration exercise time. The good side is that after exercising, Ron can plow through work quickly and without the afternoon crash in energy that others experience. Downside is that even after taking a cold shower; the post exercise sweat sometimes makes for a pungent odor.

I've run and worked with Ron and appreciated his dedication to staying in shape and his get-up and-go attitude. The one thing you can count on with Ron is that he never lets running get in the way of work and completing his task on time. So in general, Ron and his running friends tend to make a project manager's life easier.

Wiseman Wayne

Think of the Karate Master, the wise soul that can provide common sense advice because there is nothing they haven't experienced over the course of their career. These are the humble knowledgeable gurus that enjoy helping others in the organization learn about how to do things right. Wayne is a rare breed of person, whose numbers will dwindle as the new global economy forces employees to change jobs more frequently. Wayne always makes time even if busy, because he knows the importance of handing down his wisdom to others that might run the company long after he retires.

Project manager must take full advantage of someone who is not only knowledgeable, but takes pleasure in educating others on the team. These are the point guards of business, the leaders that see the whole court, can talk about opponent tendencies, strategies to succeed and loves to dish out assist so others can score. Wayne, if used correctly, makes everyone around him better and does so with a gentle smile on his face.

Knob Dick Kevin

Knob dick is a term created long ago for technical people that could never quite finish a project because they continually moved control knobs and "dicked" around trying to perfect the solution. Older people like yours truly have performed this ancient ritual trying to adjust the contrast, sharpness and color of television sets of yesteryear. There is nothing wrong with Kevin as long as a limit to the tinkering is set or the activity will go on forever. Unlike research people curing some disease and needing ample time to proof the benefits of a treatment, businesses can't afford to spin around in endless circles or run in place. Running in place is the best way to get passed by a competitor.

For the project manager, Kevin can be an asset early in development as his ability to fine tune "things" lends itself to a better product or service. Kevin has to be monitored by someone so there is an end to the fiddling and the program doesn't experience extensive delays. As with all project members, there is always a fine balance between their cost and benefit.

The End is in Sight

If this list of potential project members seems endless, in reality it probably closely mimics the truth. As each snowflake and fingerprint is one of a kind, so are project members. With each new program, the quandary of dealing with people is a never ending learning experience. I personally enjoy the challenge of putting together what is best about people when it comes to developing something new, while controlling the intended or unintended roadblocks of human nature. This constantly changing puzzle never gets old, but can certainly cause accelerated aging. In the end, there is nothing better than a team of diverse individuals coming together to create something unique.

Sweet but Clueless Cathy

These are the people who always smile, bring in homemade pies and remember everyone's birthday and are just so damn huggable. However, when it comes to getting the job done, you want to choke them, because they have no clue. For years, no one ever wanted to tell them so, because of their beautiful and caring inner person. Well businesses don't make money if all of their employees are sweet and clueless. Someone has to figure out how to get Cathy working towards tangible and profitable activities.

Okay, maybe that's a harsh assessment of Cathy, but no one wants to pay an individual who doesn't contribute because they're sweet. Don't believe many companies hire your grandmother (not to say that she isn't a beautiful person) to bake cake for upper management or make tea and toast when someone at work is sick. Well, Cathy can be an asset for a project manager when it comes to entertaining customers. People always appreciate kindness, and putting Cathy in front of customers fools them into thinking that everyone at the company is this accommodating. But if there is limited opportunity for the company to take advantage of Cathy's sweetness, it's time for her to find a new job.

Underutilized but Content Chris

Companies have these gems, the stable worker with the intelligence of a brain surgeon that is perfectly content moving pallets on the dock. Most of today's workers if asked believe that they fit in this category – working at a job that doesn't tap into the vast extent of their mad skills. If the truth be told, many of these workers also suffer from illusions of grandeur. Chris on the other hand just really enjoys the work he is doing even though his abilities can help the company in many other ways.

There is nothing wrong with having Chris on your team, since whatever job he is doing will be done without issue. However, the better project managers try to pull out some of Chris' hidden talents in order to show him how much more valuable he can be to the organization and the project. Just a word of caution, allowing Chris to use some of that hidden IQ could also cause him to go elsewhere. Project managers that want to utilize Chris and ramp up his output could lose him all together.

Cockroach Curt

These are the individuals in every company that somehow manage to survive (ala a cockroach) the continuous onslaught of downsizing in corporate America when their abilities are marginal at best. Do you sense a bit of jealousy? Absolutely, because it's almost impossible to figure out how they do it – maybe they have some magic voodoo. What's even more impressive with Curt is his ability to move to the next job just before his previous organization is cut.

Curt is usually pretty skilled at the interpersonal side of the game, golfing with managers, calling customers out of the blue to chat about the Yankees or meeting a supplier out for beers. His ability to network with the right people is excellent, but his contribution to the team is negligible. As you can guess, I'm not a big fan of Curt and have on occasion had him as a team member. Even though his value as a networker helped us, Curt's cost to the program in late deliverables made him one of my least valued team members.

Legacy Larry

Whether this person is the son or daughter of a former executive or the owner's nephew, this character is more common than you think. Even with the immense importance on human resources hiring the best and most diverse candidate, these individuals are placed in well-paying jobs and survive and flourish. Many have intelligence and talent, but fast track to higher positions because of their legacy instead of tangible results. The big problem with Larry is that his legacy and the kudos he gets from being a descendent of business royalty goes to his head and he is more comfortable being the project lead instead of a team member.

I've seen some project managers placate Larry with some mini project responsibilities to validate his royalty, but this just made things worst. Project managers rarely have a choice in terms of team composition so Larry's going to need a place. There are two ways to address Larry, find a job that goes to his strength or keep him buried in work. The good thing is that Larry usually lasts for short period of time on most projects, so the project manager doesn't have to perform this act for an extended period of time.

Tryant Tammy

Think of the female manager that decides that the best way to succeed is to act like a male tyrant even though doing so is not necessarily a formula for success. Part of diversity is show casing different ways of managing a business and people - and acting like someone else (especially the old school arrogant white male) does a disservice. Even in today's more diversified society, it is not unusual for Tammy to want to prove to the boys that she can go mano-e-mano with them. At this time in society, that approach will do little to motivate people and female executives can contribute so much more to business.

As a side, one of my close friends traveled to Japan often to deal with a partner and befriended some of their engineers. One day he was trying to help them understand situations where the slang OH SH_T could be used. After many failed attempts he had given up on teaching this classic slang. My friend returned to Japan about three months later. One of the engineers greeted him with an enthusiastic smile and told him that the Japanese finally understood the slang. His point, every time Tammy-san traveled to Japan, he and his peers would look at each other and say OH SH_T!

Now this story doesn't mean all Tammys are bad, but the business world is a lot better place without more dictators. I will say that the Tammys I've known over the years realized that being themselves instead of the dictating male served them and the company well. Many of them ended up being great project leaders.

Oblivious to the World Olsen

Many of these characters can be found in research and can walk by you or bump into you in the hallway and not even know you're there. Their mind is somewhere else and to leave this place of intellectual bliss to exchange pleasantries' is just not going to happen. These bright minds do create some extremely valuable inventions, but once in a while a "Hello" would be nice. Olsen's manager worries that one day, Olsen will be in deep thought and get run over as he crosses the street on the way to the company parking lot. The key is how to get what's in Olsen's brain in practice to make some revenue for the company.

I don't necessarily want Olsen on my team as he is much more valuable thinking up new technology in his fortress of solitude. However, as project manager, I must understand what intellectual property (IP) is available in the beginning of a project and if needed bring Olsen along to help the team. Olsen might not always know where he is walking, but he comes to life when explaining something that is near and dear to his heart. Olsen's inventions are his children and all parents like to brag about their children.

MILITARY MARK

Many ex-military can be spotted from miles away, perfectly pressed clothes, short & neat haircut, shined shoes and that erect posture that they tried to teach you in grade school. Along with this is the "do as I say without question" attitude that was embedded in them in the military. This is not bad, as long as these individuals know that businesses are about a collection of people and ideas, not a command dictatorship. However, in some situations, the strength and discipline of Mark's military background can come in handy when a project team needs a quick kick in the ass.

I've worked with and for Mark's in the past and find him to be a valuable peer. If you have ever played good cop/bad cop in business, Mark can play this role and make it Oscar worthy. On occasion in a past project, it was not unusual for me to invite Mark to a team meeting when assignments were starting to slip behind schedule. After a visit from Mark with his fresh brush cut and polished black shoes, my team always managed to catch up for fear of facing Mark again with a late assignment. Boy do I miss working with Mark.

NEVER RESPOND TO AN EMAIL NICK

In today's wired world, there are some that do not answer emails in a timely fashion or at all. However, when they send an email, it's marked priority and needs to be answered within minutes if not seconds. Failure to answer their email often leads to them stomping down to your office. Hard to figure out the wiring in their brains, unless the only things important relate to them or they only respond to a serious interrupt for support.

Don't be fooled, Nick is alive and striving in your organization. Nick's slow response might be out of survival based on the incredible amount of daily emails he receives and triages. Nevertheless, this is really not an excuse if this becomes normal behavior.

I've worked with two Nick's in my life and both were technically brilliant and overwhelmed with work/personal issues. Having them on my team was clearly a good thing, and the best way around emails was just scheduling a face-to-face meeting or picking up the phone, yes phone and talking to them live. Another approach, bringing up an item up in front of the team, as anything stated during a team meeting always took precedence.

Laid Back Lucy

Maybe these characters are children of Hippies from the 1960s and still have some good old hemp as part of their gene pool – but mellow is the word. There is never any sense of urgency, but not in a lackadaisical way, just the way they interact with the world. On the good side, these people are typically talented, but just need the proper lead-time to get to the work you want completed. Lucy never worries about work related items and her office is usually filled with homegrown vegetables or plants. However, beware if there is a Jimi Hendrix photo in her office.

Lucy is a good person, as the project manager, you just have to provide the right amount of runway in terms of lead-time or you can get very frustrated. If given ample time to perform her duties, Lucy can produce some amazing results. In addition, she can whip up a mean organic salad for the occasional team celebration. So reach out to Lucy, but pad her times a little bit in Microsoft Project to keep your sanity.

Violate the Values Vinnie

Every organization has one of these employees no matter how much the company promotes their ethics and values. They stomp on people in front of others, cuss for all to hear, make lewd comments to female co-workers and somehow avoid the values axe. If you end up on the wrong side of these individuals, watch out even if you somehow become friends with them, always remember that Pit Bulls turn on their owners. Vinnie is a corporate bully and even though schools have done a lot to prevent bullying in the classroom, corporations have no clue.

There is no need for Vinnie – period, and how he survives is a question asked by pretty much everyone in the organization. To add insult to injury, many times Vinnie is picked to lead the project team that does nothing to promote team unity and success. Bottom-line, there is no need for a bully of any kind in a company or on a project. If you get a chance, send Vinnie to where he belongs, back on the mean streets to fend for himself.

Finally done, thanks

Even though these personalities have been placed into tight and limited characteristics, in life, we are all a combination of many traits that make us unique. The names and character traits can go on and on, never ending, but the job of a project manager is finite, get the job done on-time, on-budget and to the delight of shareholders. Just as there is a partner for every man and woman, there is a place on a project team for every employee. The project manager has to take the individuals just mentioned and the many others and use their strengths to make the company money. It's simple enough, just ask any manager.

Conclusion

Here is an obvious statement- people are going to be the key to either a successful or an unsuccessful project – duh! Well if this is so obvious, why do so many projects fail due to having the wrong people with wrong skills and wrong attitudes on them? In construction, they don't hire electricians to install plumbing and vice versa. Much of this can be contributed to the lack of a clear people strategy or comprehensive plan before a project officially gets started.

With The PMPs, it is critical to look at each of these unique ingredients to a successful program in the beginning, which includes people. If you realize after a project has started that your team doesn't possess the skills for success, don't lose hope, it's your job to right the ship. Football teams have pre-season to prepare the team through rigorous exercise, drills, video study, etc. to make their way to the Super Bowl. However, the selection of players could have been in the works for years and continues throughout the season. Business projects don't have this long-term view of comprising teams. They don't have the equivalent of the NFL Combines to measure how well the person will fit into a team including their 40 yard dash time to the coffee pot.

Many times the success of a project team can be pure luck as if the Gods shined down on the selection process and the right people with the right skills and attitudes all came together for a single purpose. However there are other times where excellent people failed to succeed as it took the length of the project to understand how best to meld their talents.

In a perfect world, all project teams would be made up of Sam, Elisabeth, Chris, Wayne, Kirk and Lucy. Instead, project managers are handed Vinnie, Betty, Mike, Mary, Cathy and Larry. Alternatively, a combination of all of the aforementioned people composes the team and one person, the project manager is left to transform them into a winning unit. In the end, the most rewarding part of a project manager's job is managing people and their unique talents.

Parts

All project managers focus on the main driver of projects, developing the new or innovative product or service with a zeal that would make a vulture at a slaughter house envious. This is just human nature, to keep your eyes on what is most risky of any project, since the greater the risk, the greater the need for oversight. No one wants to let the development effort go on without constant attention. Taking your eyes off the project for hours or even days can lead to disastrous results downstream.

But lurking below the surface of most projects is another creature; one that if left unattended can damage your project more than Godzilla loose in Japan. Its parts, yes those items that when put together and tested actually lead to revenue. As much as design problems tend to keep project managers up at night, parts can cause even greater harm to a good night's sleep. With some parts having lead-times in months or even years, failure to attend to these in the very beginning of a project can sink the program, with little room for recovery. Just imagine Apple coming out with the next version of the iPad and forgetting to order the displays in adequate quantities for demand. Well Apple is certainly not going to let this happen, however, many other companies find themselves in this position all the time, with too much focus on development and too little on the supply chain.

The Stealthy Critical Path

Project managers most often have to keep their eyes on the main driver of the project- the risky and costly development of a new product or technology. The majority of their time is spent keeping track of getting that first unit, the dreaded Engineering Model (EM) working so that the next steps in the process such as quality testing and then production can occur. After all, the majority of the dollars and time are tied up in those expensive engineers and scientist that have the task of pouring their brain power into a profitable product or service.

Many project managers excel at being able to drive this group of disparate characters into a money making entity while adding another notch to the project manager belt. Just as Boy Scouts take pleasure in showing off their merit badges, project managers feel the same about projects. But there is something else that has destroyed many projects – the lead-time of raw material, parts, subassemblies and tooling needed to eventually produce the product. Most engineers don't think about lead-time since there's nothing technical or sexy about parts, what counts is seeing tangible progress on the design. This goes back to the time of Dr. Frankenstein and his delight at seeing his monster, neck bolts and all come to life. Dr. Frankenstein was certainly not worried about the lead-time of more human parts to make his next creature.

Adding insult to injury for an engineer's lack of concern for part lead-time is the Technical Sales representatives - the crack dealer for engineers. They don't climb down chimneys, but frequently appear and give out samples of the latest FPGA or IC. Engineers don't shop at The Dollar Store, but look forward to the day when their local electronics salesman shows up bearing gifts. This all too often planned encounter never helps the project manager as assumptions to the real lead-time for the part are poisoned by the immediacy of delivery by the representative. Once the engineer is hooked on this part, it's a sure bet that the next design will include one of these items - the one that was handed to him or her by the representative.

With the team's urgent need to get parts, really anything readily available in small quantities for the EM, the overall lead-time of production parts gets masked or is grossly underestimated. Therefore the potential part lead-time has on the overall project is just exacerbated. Reason, it was so easy to get parts for EM, how hard can it be to procure parts again? It's one thing to get parts for one unit, a completely different undertaking to prepare for full production. Given the current recession and companies pulling back on production, on hand inventory is lean, and lead-times on some electrical components can be anywhere from one month to one year. Ignoring this issue while driving towards a working unit will do nothing but prepare the project manager for failure or sleepiness nights trying to figure out how to get production parts.

In the Beginning

What will truly separate the good from the great project leaders is asking the following question at the very beginning of the project – "What is supply chain's longest lead-time item(s)?" Okay answering this might be difficult in the beginning as engineers hide in their cubicles and design away, like Santa's elves at Christmas, but the question should be one of the key action items to address as soon as feasible. It's the project manager's job to oversee all lead-times including parts, subassemblies, test and production equipment. Without an understanding of these lead-times, there will be an additional to the truisms of life: death, taxes, and late projects due to a lack of parts.

Many times, parts and lead-times are not front-and-center as the rush to get a working engineering model is the sole focus in the beginning and all short-cuts are taken to meet this end goal. Therefore, the perceived time to actually build production units is masked as purchasing is kept outside the Design Palace until the arrival of the EM. However, once the truth is known about lead-times, everyone points to purchasing to somehow improve delivery of items they have no control over.

What must be done is to understand those critical long lead-items as early as possible. With this information, the project manager can put place holders in the schedule dictating when a portion of a design or even a specific function needs to be frozen so parts can be ordered. Is there risk, yes, but without taking intelligent risks, there will be no on-time schedule as the production departments and all of its operators, support staff and equipment, wait for that last part to arrive.

On the good side, many times the number of critical, expensive and long-lead items can be counted on both hands without need of your feet. This must be formulated early and shared with management so they are aware of the situation before the project manager moves ahead and buys them. Then again, what do managers know about project management since many of them have not been in the trenches?

History and MRP

Everyone uses historical information to make decisions in the present, and this theory applies to the supply chain and parts procurement. [Even an implementation of an MRP system like SAP requires hardware.] What better gauge for long-lead and/or problem parts than what is happening now when things are fresh and pure. After all, crime scene investigators want the scene unencumbered, and getting to the root cause of long-lead times and poor quality parts is best addressed as soon as it happens. Purchasing doesn't like to deal with problems any more than the project team and is more than willing to work on providing high quality and shorter lead-time parts to production.

All MRP systems have data on standard lead-times, actual lead-times, parts received, accepted, rejected, etc. Once an engineer starts to design his or her board, cable, enclosure, etc. the availability of information from MRP can and should be used to understand what parts could be an issue in terms of launching a product or service. Yes parts (or equipment) are every bit as important is developing a service as they are for delivering a product.

The question is how the engineer gets this information; after all, the Technical Sales representatives are all about the solution, that technical high when the cool new electronic device is handed out to the engineer. How does the project manager force his engineers to visit the purchasing group to utilize their expertise? Have a representative from supply chain on the project team and make it clear to all that his assistance will be used to review all bill of materials to help improve the parts procurement process to support the project schedule.

The beauty of having supply chain involved upfront is that what the MRP system cannot provide in terms of critical parts/supplier information, the purchasing agent can provide based on experience and hard knocks. They have lived through the top-down beatings when one item is late from a supplier and millions of dollars of lost revenue are hanging in the balance. Especially towards year-end when management starts to scrutinize all metrics that put bonus money into their pockets.

Use the Parts Dealers and Suppliers

It's easy to pick on Technical Sales representatives, but the good ones do provide a benefit, keeping engineering and supply chain abreast of the latest technology. In the past, this was less important as technology changed more slowly, but nowadays, with products changing quarterly, it is imperative for a company's success to stay on top of the fast moving technology wave. The other benefit is that many times, there are application engineers that support these representatives who can help with the actual design.

There is always a love hate relationship with suppliers, but good project managers know that a premier supply base makes their life much easier. If project managers could select their team they would go the route of the Miami Heat with players like Lebron James, Dwayne Wade and Chris Bosch. Why wouldn't the project manager want to apply this same principle to everyone person (and partner) that touches his project. This is my approach, since I don't like to lose, and would never settle for second best in terms of choosing my team.

Suppliers understand not only their own quality and lead-time, but issues with second, third tier suppliers that can be critical information in the hands of the project manager. Many individuals fail to realize that lead-time is the time it takes to get enough good parts to build revenue producing products. If a supplier has a 5% failure rate and the normal lead-time is 10 weeks, a percentage of parts could take up to 20 weeks to receive. Try explaining that to your manager, revenue for 5% of your sales will be delayed for 5 months.

If true supplier partnerships are in place, purchasing can provide the linkage between engineering and suppliers early on to start addressing parts for both engineering and production. This enables each partner to share prior experiences (history), current and future state of parts to provide all options to develop and sell a new product. Great suppler partnerships go to the extent of sharing product roadmaps with each other, so when a company starts the next generation development project, the supplier has already invested R&D. R&D to develop a unique part, subassembly, material or production equipment that can make both companies money.

What's this have to do with lead-time, everything, because the ability of the supplier to start the R&D process well in advance of the need reduces the overall schedule. I've had experiences with ASIC development that can sometimes take years to design, test and produce. If this is done prior to the start of a program, the only lead-time is due to the factory – as the design, process and testing to get working ASICs are of thing of the past.

Design Practices

What's that saying my parents always preached when I started guitar lessons, oh yeah "Practice makes Perfect." When it comes to company success, this cliché is addressed by implementing process after process hoping that after years of fine tuning the process, if it's followed, success is the only outcome. If a design has worked in the past, why not use as many parts from this as possible, even though designers live for the next great invention.

In many successful companies, there is a push to use existing components that have been proven on previous designs. The biggest issue with parts from a scheduling standpoint is lead-time, followed by variation in quality and price that can change over time. The practice of using existing parts that are used on current products helps take the guess work out of the supply chain. There are always going to be parts problems, it's a natural occurring phenomenon in business, however, why not learn from the past transgressions and go with what's already working.

The other benefit besides understanding the heritage of existing parts is the savings in the day-to-day activities needed to deliver the next great technology. For example, circuit board layout depends on a CAD system with part libraries that show pin definition, geometry and other critical parameters so that a board can be both manufacturable and testable.

Having years of experience centered around electronics, I can tell you that many programs take a schedule hit from something as simple as using a new connector. There's just something mystical about figuring out which location is Pin 1 from either the top or bottom view of a connector that has caused trash piles of useless boards. Using an existing connector allows the program to not suffer this fate as others in the past have paved the way from their failures.

The project manager's world is a balancing act especially as it relates to the ongoing battle to manage the supply chain. Certainly using parts already in use is a good practice, it must be balanced with the foresight to design in the latest technology to stay ahead of the competition. One way this can be done without impacting a project schedule is with advance development or R&D. The strategy to experiment with new technology prior to the start of a formal project allows the corporation to work through and around technical issues, while the supply chain gains experience. This is a win-win situation, new parts can be vetted, supply chain can setup the appropriate suppliers and the critical nature of ordering parts can be managed more efficiently.

Separate Schedule

The ability to order and manage the procurement activities for a major program necessitates the same level of focus on activities and resources that go into the overall project schedule. If you don't think so, just imagine being solely responsible for all the raw material, parts, subassemblies and equipment required to build Boeing 747-8s or Airbus A380s and an overarching schedule makes sense, in fact is a necessity.

Too often, project managers believe that a handful of line items on a schedule can encompass the complexity of ordering all of the parts needed to launch a product, but this often leads to a late time-to-market. In defense of project managers, their expertise is not in supply chain, however, as the leader, the buck stops with them. The practice of having deputy project managers or Integrated Project Team leads helps offsets the project manager's workload - the supply chain portion of the project needs a leader as well.

The supply chain leader must drive the design team to provide the most critical element of all – the Bill of Materials (BOM). As mentioned earlier, a handful of line items in a schedule cannot fully address the tasks needed to deliver parts, tooling, equipment and packaging material. There must be a separate schedule or better yet, a BOM that can be released into an MRP system that has the power to track inventory needed to build production units.

There is no way that a project schedule can justly capture the critical path for parts, however, an MRP system can slice and dice this in so many ways it will make your head spin. As with most projects, the focus on getting the first unit built overshadows the need to develop a comprehensive BOM. What happens next, once design is happy with what has been created in their lab, time is spent on the painful process of documenting the formula or parts list for the product. The big problem here – part lead-time has been consumed during the design process and once parts are truly placed on purchase orders, they're already late.

The beauty of any schedule is that it tells you the critical path(s) and allows what-if scenarios to shorten or eliminate those Johnny come lately activities. Parts are and always will be an item on the critical path, what better way to manage this problem than to see it in all of its glory. Once it is visible and in plain sight, the project manager or part Integrated Project Team (IPT) leader can determine who best to drive the supply chain to success.

Take a Chance – Roll the Dice

Yes, in business you can take a chance, roll the dice and win without being in Vegas. Part of taking ownership of the supply chain and parts is to use time to your advantage. If the mantra of starting a new business is location, location, location, then the one for parts is lead-time, lead-time, and lead-time. Project schedules have a finite life, there is no such thing as a never ending project with an endless supply of money unless you are the US Government. Given the fact that time machines are only real in movies starring Michael J. Fox, someone has to be willing to order parts in the very beginning of a project to use all of the available lead-time.

Projects managers can do their best to add drop dead final design dates to force the ordering of long lead parts as mentioned earlier, but in reality, this is a game of chance. At some point in time, the dice will have to be rolled. The best evidence to save your backside is to take a chance with facts - understand the lead-time of critical parts, their cost, the probability of scrapping them and the potential benefit to drive schedule and revenue. Sometimes this is a no brainer exercise, other times it could be a career limiting move.

However, hindsight is 20/20 and project managers are damned if they do and damned if they don't buy parts early. If you guess right and keep your schedule on task, few recognize your brilliance, guess wrong and scrap parts that impact management bonuses and the world will know of your futility. For me, I'm taking the chance and ordering the parts, there's no way I want to give the competition a leg up and hurt the shareholders for what can be a easily managed game of chance. So, go ahead and order the long-lead parts and ask for permission later!

Conclusion

If you talk to any project manager that has led a product development team, parts and "releasable software" are probably the last items they usually need to start making money. Grant it that marketing and their intelligence (or lack thereof) can impact both with last second feature creep, however, project managers are not allowed excuses, they must "Just Do It." The benefit of having a better shot at managing the supply chain is that in many cases, previous lessons learned can be applied, thanks in part to the knowledge of the supply chain organization from past projects.

Given the need to best utilize lead-time, parts and their impact on the critical path must be foremost on the mind of project managers. Just as project managers have to deal with ongoing design problems, specification changes, and competitive actions, so do suppliers that support them. Giving the suppliers the ability to start early; provides them with the same level of flexibility to manage their unique problems in parallel with the project.

Successful project managers don't lose sleep over parts, they use premier suppliers, knowledge from supply chain, past success with parts, and the mindset of a poker player to align the parts with the need date in the factory. They know not to get fooled by how easy it is to order parts for that first engineering model and keep keen eye on the revenue producing parts.

Parts as well as everything else associated with a project are about managing variation, the less there are the more time and flexibility is at the project manager's disposal. However, many times parts and their lead-time have a mind of their own, and can drive even a seasoned project manager to drink, heavily. Rest assured that there are plenty of ways to compensate for supply chain problems along the path, but one of the best is to start early and review often. After all parts are just parts.

Process

I love to watch the Food Network; there is nothing better than a world class chef preparing gourmet meals in front of your eyes. Meals that look so delicious you could almost taste them through the high definition LCD television screen. The ease with which these chefs create culinary masterpieces appears to take nothing more than mixing the right ingredients, placing them on the stove or in the oven and minutes or hours later having something that will amaze your palette. But what the watcher fails to understand is that prior to the show and years before hand, the chef developed his or her skills through practice and process, or vice versa.

All things require a process; you have to first put the clothes in the laundry before adding detergent and setting the time and temperature. To do these things in another order could cause severe consequences. For example, mixing red clothes with white, adding bleach and setting the temperature to hot is a horrible process for cleaning whites.

Processes help to guide humans, assuming they are paying attention, as a way to minimize our inherent ability to screw things up. It's a "Don't Vary" checklist that will lead to good results most of the time. Processes force people to follow a well-defined set of steps that have been proven to yield excellent results over time. The question for project managers; what's the right level of processes to guide the team, while allowing them to be creative?

The Process Conundrum
Goldilocks and The Three Bears

Introduction

In the early 90s the rage was re-engineering the concept created by Michael Hammer that gave corporations the tools to reengineering business processes to solely focus on the customer, without wasting precious resources in the process. Hammer and James Champy co-wrote *Re-engineering the Corporation* in 1993 which became a bestseller. Hammer traveled the country and lectured on his concept with audiences showing great enthusiasm for the freedom to attack wasteful business processes and turn them into a panacea of process bliss.

Back then, I was assigned to represent our operations group on a major reengineering project whose goal was to redesign the whole order to installation process for high-end medical imaging equipment. After one very successful reengineering project, our management team decided that two times would be a charm. However, the depth and breadth of this effort was tenfold the previous. On the bright side, being someone that always thought things could be done better and falling victim to Hammer's clean sheet of paper mantra to start anew with streamlined processes; I was ready to dive into the project.

After less than a month, the process to implement the reengineering process was more than I could take. In fact, the process to fix the process overtook the whole project. The manager in charge basically took the creativity out of reengineering by throwing process after process into the project to repress any and all imagination. And the conundrum was born, do processes help project managers succeed or do they doom projects to failure? Today's project managers are faced with processes galore, ISO

9001:2000 (general industry), AS9100 (aerospace), ISO/TS 16949:2002 (automotive), FDA regulations, and the tried and true Phases and Gates.

On the other hand, there are still many small companies that have not moved into process hell; and flip side of the conundrum, are you better off with too few versus too many processes in terms of successfully commercializing a product? Having experienced both sides of this argument leads me to the following: "It Depends."

Certainly no one will argue that developing a quality process per ISO 9001:2000 to mitigate or eliminate problems is a bad idea. The problem is how to satisfy two unique needs, one to launch new products quickly to drive revenue and the other to do so by adhering to both internal business processes and external standards at the same time. Processes are guidelines to make sure that all the appropriate ingredients to success have been considered, but good ones leave room for people to do their best within certain parameters. With too few processes, people have no guiding light, with too many; they're in the dark going nowhere fast. In my experience, the corporations with too many processes were the result of those infrequent problems that delayed a project and the immediate reaction was to add another process step instead of addressing the root-cause.

To Process or Not to Process

Yes Shakespeare has a place in industry when it comes to the question of business processes, do you really need processes? And if so, taking from Goldilocks and the Three Bears, when are they too hot, too cold or just right. Let's cut to the chase, everything we do in some way is a process, from how we go about doing the laundry (remember separate the colors from the white – as red clothes and white produce pink) to preparing a meal for the holidays. We are just so used to doing this that it becomes part of who we are and the notion of following a process is not top of mind.

Still even without a formal laundry process, there are occasions when something red sneaks into the whites. This has happened to me on rare occasions, producing a load of clean pink clothing. Try telling your wife at that moment you were only helping out and a pre-laundry process with a check list starts to make sense. The decision is how to balance the amount of business processes with trust in your human assets in order to have the right mix for success? Too much process kills creativity, too little leaves too much to people, so you are between two extremes. Project zombies on one hand to the businesses equivalent of Animal House on the other.

As mentioned before, even if a business owner did not believe in overly burdening his or her employees with processes, customers' expectations many times are to purchase from companies that are ISO 9001:2000 certified, which requires well-defined business "processes." The easy answer to the process conundrum is simple, put enough processes in place to guide the project team, while leaving room for their ingenuity to determine the best path. Nothing is ever that easy since people are responsible for developing of modifying processes. Choosing the right processes let alone the amount of each is a very difficult challenge.

Case in point, at one of the companies I worked for it took me four weeks to release a drawing. Part of this was my lack of training to perform the required process steps, but even after getting help and downloading the procedure within a day, the rest was a painful swirl. The swirl consisted of checks and counter checks, proper corporate and program document structure, program specific part numbers, reviews both from release and program teams, and finally getting the eight team members to approve. This was way too much process – too complicated, time consuming and expensive!

Phases and Gates

Let me start by confessing that I'm not a big believer in convoluted processes. Why, because I believe in some warped sense that having overly detailed processes means that management doesn't trust that I know what to do. Or if left alone without a formal process I will cause the financial ruin of the company. Convoluted processes as far as I'm concerned are for those in industry that still need their hands held and lunch made by their mommies. Okay, my opinion is a little much; however, the Lockheed SR-71 Blackbird plan that traveled at Mach 3+ and performed strategic reconnaissance was developed by a group of rogue engineers under the direction of Clarence "Kelly" Johnson. Given the huge advancement in technology to produce the SR-71, there was no way a process would have helped guide this talented group, it was Johnson who guided them.

Well, over the years, my allergy to processes has been tempered and the first step in the journey was Phases and Gates. A long time ago, while working for an unnamed Fortune 500 Company, I was introduced to a well thought out formal process with checklists for commercializing new products. The process had Concept, Technical Feasibility, Design, Launch and Discontinuance Phases with a Gate review at the end of each for approval to proceed. Today many companies have a Phase and Gate process, but with ours there was the ability to utilize the creativity of team members, trusting that things would get done within the bounding box of the phase.

In essence, we were allowed to follow a recipe and take steps to improve the quality of the end product through the collective experience of the team. The other positive benefit of the process was that everyone who worked on projects had to follow this process, so it became ingrained in our culture. Another benefit, the team, not management, committed to a launch date after passing through the technical feasibility gate, putting the ownership for success squarely on the team. Sure there were management challenges posed to meet critical dates such as units available for critical trade shows, but over time, the teams managed to do so without direction.

The gate review process was very simple, send all appropriate documents to the management team a week in advance, choose a time slot (as all meetings were held Tuesdays from 8AM through noon) and present issues only. Again, the managers trusted that the team was meeting the key deliverables and had no intention of going through reams of detail; they wanted to know what could cause a cost overrun and or schedule delay. There was also a bounding box related to how well the project was tracking to the spend profile on project and business case. Failure in either could cause your show to be cancelled. Last, each team had to present all issues with plans to solve them including cost incurred and potential program delays.

Getting the whole Phases and Gates to this level of maturity took around 5 years. But the end result was well worth the effort. Our team used this to great success, launching two major new follow-on products in 9 months after technical feasibility. However, even with this process, other teams failed to stay out of the project doghouse. Bottom-line, we had the right amount of process to guide us, the ownership to set a launch date and the freedom to use our experience and skills to achieve the goal.

Inventiveness

A mosaic is creating a decorative piece of art from assembling small pieces of colored stone, glass and other materials. The skills of the master mason are put to the test and even if outsiders might not be aware, there is a process that guides the effort. Imagine if the mason was subject to processes that required paper work to document the size, shape, color and weight of each piece, how he or she selected the raw piece of stone or glass and all the steps necessary to chip away at the larger piece to get just the right one for that location within the mosaic? Well, not only would this stifle the imagination of the artist and take more time, it would probably lead to flying tools towards the person who created the processes.

Now take the same example above and do the complete opposite with pharmaceutical companies, telling them that there is no need to document anything when it comes to producing drugs. That's right, no need to understand chemical makeup of the various ingredients required, no need to measure process parameters during batch production and no need to monitor how the drugs proceed from batch production into pill and or caplet form. Not sure if you would get any takers for these drugs, even though it would cost a lot less to produce them. The cost savings in manufacturing would be overwhelmed by the price to settle the various lawsuits.

What's the point; the best processes are those that ensure the safety of the end consumer, while fostering the creativity of the team. Think controlled creative chaos! People are unique animals that can think and develop new ideas, many of them that can lead to more revenue for a company. What's the point of paying someone a salary of $80,000 and getting $20,000 worth of revenue out of them? The better approach is to foster inventiveness to turn that $80,000 salary into a $320,000 in new revenue.

Many organizations are compelled to corral their employees under the guise of efficiency, not trusting them to behave in the best interest of the company; that just leads to a very inefficient team. However there is a careful and delicate balancing act, because too much of a good thing (as in freedom) can lead two extreme outcomes – success or failure. Project managers must be able to stay within the bounds of the processes, yet provide enough freedom to team members to extract the most value.

Trust

If you've ever been a part of a small entrepreneurial company that eventually morphs into a larger one, employee growth is always followed by process growth. When there are only two or three employees who decide to start a company, the one key ingredient that they have is trust in each other. Taking a risk to venture out on your own into the corporate jungle requires a foundation of trust when things get difficult. As the company grows, strangers, at least in the eyes of the founders are brought in to spend money directly out of their pockets. Project managers hand money to the project team, wondering whether or not the price we are paying is fair and if the team is committed and trustworthy. If there was total trust, there would be no need for someone to guide the team; everyone would somehow behave in the best interest of the company.

It is one thing to put in place processes to guide the company, making sure that nothing will fall through the cracks, sort of a how-to-do checklist. But coming up with a corporate process on how to go to the bathroom and wash your hands afterwards is a bit ridiculous, even though the later, is available nowadays. I once worked at a place that not only told you to flush, but how to wash your hands and to clean water that spilled on the counter of the bathroom sink. Yes, some people are slobs, but really. There are just some things that common sense should handle without the need for a process. However, humans have flaws, which lead to problems, which frustrate customers and deplete corporate profits. Solution - add another process or process step!

Trust in people would allow management to always analyze the problem first before jumping to conclusions. Phases and Gates is a process for new product development, but in between the phases, it is free rein for the team to steer the project to the next milestone. Managers must trust in the skills and experience of the team between the phases to perform their job in a way that will yield a money making product or service. To withhold trust and place process obstacles in the way of the team goes counter to what the goal of the firm must be – maximizing shareholder wealth. Shareholders would be very unhappy to witness a process maze that does nothing more than shred their dividend check.

Trust is a great motivator, something that pays back tenfold to the manager that provides this gift. However, trust in individuals doing their best as part of a team doesn't eliminate people's tendency to make mistakes. Processes that can help prevent mistakes while promoting trust are the perfect balance that companies must achieve in order to effectively get the most out of their investments.

Root Cause

"Root Cause" is probably the most over used statement in business and an oxymoron along the lines of marketing intelligence. Yes, these are the words first spoken by lean six sigma babies. If I had a penny for every time I heard someone say, "We have to get to the root cause of the problem" my bank account would make Bill Gates and Warren Buffet envious. Why bring this up, because many times the reason behind the latest and greatest process or process improvement can be directly tied to mitigating the root cause of a problem, especially as it relates to a customer issue. However in reality and in my over 30 years of industry experience, I can count only a handful, less than five times, that the true root cause of a problem was actually solved.

Many other times, what was perceived as the root cause led to another process that ended up costing the shareholders money even though the addition of the process was done with good intentions. And in this sue happy culture; a company would rather put in some process that can be used as a defense in court than permanently fixing an issue. Back to common sense, to the elderly lady that sued McDonald's for burns suffered from spilling hot coffee on her lap, I say duh, coffee is hot; she certainly waited to drink the beverage or took tiny sips so as not to burn her mouth many times before this incident. Spilling hot coffee on her lap should cause some immense pain. Putting a label that states "Careful, the beverage you're about to consume is extremely hot" doesn't prevent another accidental spill.

In this lady's defense, the root cause could have been a poorly designed or manufactured cup and lid that did not close completely. After all, there are no fail safe mechanisms in place during most coffee sales. As a frequent consumer of Tim Horton's and Starbucks coffee and a shareholder in both places, Tim Horton's cups are poor at best and leak quite frequently, leading yours truly to be very careful about consuming coffee, especially when driving. Do I believe the lid will completely come off? No, but there is always a chance of perimeter leakage that happens on occasion?

Let's assume that going forward and after reading this article all true root causes will be understood and process changes will permanently mitigate this problem from happening again. Well, I have another suggestion on the subject of root cause and process change - instead of adding another process step or creating a new one, how about reviewing the whole chain to see if it can be improved? Having worked in both medical and government corporations, the tendency is not to change something that is perceived to be working. Process after process or change after change is just added without ever stepping back and viewing the whole. What happens, there are multiple steps in the process that ask for the same information from different departments and no one realizes the tremendous waste of time and money.

Conclusion

Back to Goldilocks and the Three Bears and how it relates to the process conundrum, companies can have too few, too many or just the right amount of processes to run a profitable business. The great companies use processes as means to ensure that products and services delivered to customers exceeds their expectation. Project leaders must weave their way through the maze of processes and somehow balance them in a way that delights the end customer. They also must know that certain processes cannot be changed, for example adhering to Sarbanes-Oxley Act for accounting reforms.

Processes are yet another tool at the disposal of the project manager; they can hold hard and fast and use processes to force behavior without being the bad guy or girl. Or push the boundaries of processes to showcase independence and "We are in this together to succeed" bravado. Processes such as Phases and Gates provide both a tool to hold the team accountable and an avenue to use their creativity to achieve the end goal. In today's business world project managers will always have to deal with the good, bad and ugly that is business processes. The goal is to use the processes and not have them use you.

Rational Rhythm and Project Management
How to Stay in "Project Shape"

Introduction

Ask any professional athlete, musician, or fighter pilot what it takes to stay on top of their game and all will emphatically answer – "Practice makes perfect!" The best example of this is Tiger Woods, the former # 1 ranked Golfer in the world – before his current break from the tour. He started training under the strict tutelage of his father while still in diapers and the result bodes well for those overbearing parents. [Yes, those parents that put plastic golf clubs into the hands of infants too young to crawl, let alone walk.] Not only do professionals spend hours practicing their craft, but an equal number of hours fine tuning their body through physical and mental exercise, nutrition and meditation.

Tiger lifts weight, runs, eats green leafy vegetables (I think) and listens to soothing music while being given a deep tissue message. Then he heads to the driving range to practice his swing before playing a round of golf. There is no end to his drive for perfection. Even after taking time off due to some off the course distractions, Tiger still manages to come back to the Masters and competes like he never left.

What does this have to do with project management? Everything, because project manager's and team members, rarely, if at all, take time to stay in "Project Shape." Project Shape is when a team is solely focused on shareholder wealth and can easily tackle an infinite number of tasks in a timely fashion, all while overcoming roadblocks that get in their way. It is a single-minded collection of diverse people that meshes so effectively that victory is almost certain. It goes beyond the capability of each member to what they can accomplish as a single entity.

The problem with Project Shape is that it is fleeting, a paucity even for the most seasoned project management professionals. Why? Because as soon as the project ends, individuals saunter off into the functional abyss to be reconditioned into the existing organizational culture. No longer are they part of a team, but rather a functional headcount trying to acclimate back into their group. Individuals that were once members of a highly flexible and skilled team, get boxed into a role that doesn't come close to fully utilizing what was once a talented project athlete.

In order to keep teams in Project Shape, we need a way to train them even when there're no major projects on the horizon. This is where rational rhythm can be used to keep the team in peak condition. Rational rhythm is the business equivalent of war games. It's a corporate strategy focused on enabling teams to maintain or enhance their skills that has a clear financial benefit to the company. This does not mean team pilates or yoga classes or time with Bruce Harmon to improve their golf swing; it means an investment that relates to tangible revenue-bearing business skills. These are skills, that if unused for an extended period, will get rusty and cost a great deal of money to get back in order to support a project.

The rational – is that there is a cost to disbanding a project team that completed a complicated project – such as launching a new satellite. Moreover, that it's going to take funds that CEOs will find hard to justify in today's economy. The rhythm – is what can be done to keep the team in "Project Shape" – during the inevitable periods of downtime between projects. Knowing that the saying; "Pay me now or pay me later" will more than substantiate the pay me now strategy, provides a solid business reason for staying in "Project Shape."

"Rational" – part of the strategy based on cost/benefit, an integrated part of culture, rationing money to enhance or maintain team skills

"Rhythm" – to keep skills sharp, planned team practices (corporate war games) to overcome learning curve inertia, timed to address project downtime

This extends well beyond the team to suppliers of raw materials, components, sub-assemblies and equipment especially as it relates to very complicated and unique projects. Given the complexity of the task, it is critical not to lose the technical skills that have been perfected over time. Starting and stopping these types of projects are very costly, due to the learning curve inertia that has to be overcome in terms of skills. Pilots can practice on a simulator for months, but there is nothing that can substitute for actually flying a plane. Rational rhythm is a strategic plan that allows project teams to fly short distances in anticipation of a cross-continental trip.

The problem project teams run into is that shareholders, and therefore, businesses focus on short-term profit. It makes business sense that when a project ends, all the functional players return to their designated corner or even worse get caught up in a downsizing. When the next project starts, the financial business case is completed in a vacuum and doesn't consider the skills lost from the previous project. Therefore, the new project takes 2 to 3 years to complete because the team requires 6 months to 1 year to get back into "Project Shape." If the company thought long-term and implemented a strategy to allocate dollars that minimized learning inertia, significant cost and time could be saved from future projects.

What can be done to institute war game exercises into the fabric of corporations to keep teams in "Project Shape"? After all, each military branch plans for these out of necessity. Companies can tackle this issue with a strategy that includes a roadmap for products (product family plans), a plan for human resources around critical competitive skills, institutional learning for continuous improvement and nurturing the corporate culture around winning project teams. Any one of these alone will not be the solution, they must be done together to reinforce the good, improve the so-so, and mitigate the bad. Companies must introduce products ala Apple; hire people ala Google and institutional learning ala The Army. Now that's a recipe for success!

Roadmap - Product Family Plans

Apple is a great example of a company that creates a product roadmap and consistently delivers upon their plans with creative leading-edge products. All three of my children are proud owners of the iPod Touch and anxious for the next version to be released. Even my wife now owns an iPod Nano. The product roadmap is like the PGA tour. It consist of different products that the company plans to develop over a specified time horizon. The PGA tour offers golf tournaments, playoffs and a championship at different courses from January to September. Playing golf on a consistent basis keeps golfers at the top of their game, while developing one product after another keeps project teams vital. There is no downtime or overcoming learning inertia, just the next challenge. The added benefit is that learning from the previous program is fresh and can be acted upon immediately.

I've seen roadmaps work well in a few companies and can attest to the positive impact they've had on keeping teams in "Project Shape." Downtime was short, consisting of a lively team celebration with food, alcoholic beverages and custom team golf shirts between challenges. (The best parties were when marketing picked up the tab!) Then it was back to work, leveraging the good, avoiding the same mistakes and keeping marketing from the dreaded feature creep. Each new product had its unique rhythm, but the band (the team) was well rehearsed from before and picking up the new song (the project) was not that difficult. The other benefit of a strategic roadmap is that teams know where they've been, where they're going and the requisite journey to get to the end.

Roadmaps that support staying in shape must consist of the following: future product plans, technologies necessary for these products, and most importantly human resources' vital to create the technology and develop the products. Many roadmaps focus only on products and technologies with no regard for the actual people needed to make the strategy a reality. Human resources must include both technical and project management skills with an eye on the future. Just as NFL teams draft new players each April to continue to replenish talent, corporations need to follow the same strategy to keep in rhythm.

However, being in shape doesn't always translate into success because teams depend on other functions, internal processes, outside suppliers and market dynamics. There are always variables that teams face that cannot be controlled. Just look at the sub-prime lending crisis and the impact it has had on the global economy. I don't care if your team is in Lance Armstrong Tour De France shape, trying to succeed as a team is always difficult. Even Lance Armstrong cannot be successful without a supporting cast. Project teams require the same, individuals that bring unique skills to help improve their chances for success.

Human Resources – Skills

Let's face facts, not everybody is cut out to be on a project team and even fewer are qualified to lead them. Yet, I've never been part of any company that plans for project leaders with the same zeal as the over-priced managers that negotiate golden parachutes and huge bonuses. Can you say "AIG"? Donald Trump should write a book on selecting project managers and teams and charge millions in consulting fees. After all, the winner of The Apprentice is the person that can best manage a project and people regardless of the task. One key element of The Apprentice is that the contestants come to the table with mad skills. Trump stacks the deck by looking through hundreds of potential applicants for the apprentice participants. However, not many companies go to these lengths to bring in people who can effectively manage projects.

What companies need is a human resource strategy to hire employees with superb skills and provide them training to constantly augment their skills. Now, before heading off into some complex psychological research into team dynamics like perfectly matching people based on the Myers-Briggs Type Indicator, why not start by looking throughout the company for those atypical, yet successful teams. These are actual examples to learn from in terms of team dynamics, skills, experience, functional interactions, processes, specifications, suppliers, management direction, etc.

It could be as simple as setting up focus groups to ask team members what were the keys to success. Part of interviewing these teams helps determine what drove success and how this can be leveraged for other programs. These focus groups must be run by someone in human resources to aid them in understanding what to look for in future candidates. For example, can a selection process for hiring be developed based on what constituted successful teams? Is there a certain set of team experiences that led them to success?

Human resource's plays a key role in drafting individuals for the company, since there must be some who possess leadership skills, others with engineering skills and so on. There must be an ongoing dialogue with the various departments in order to formulate an employee plan required to successfully develop products in the roadmap. After all, people are always going to be the company's greatest asset (and liability). Google is one of few companies that have this down to a science. Potential candidates are put through a litany of screening processes in order to match them perfectly to the Google culture. Google hires extremely intelligent, creative, hard-working team players that tackle challenging projects that make life easier. Well, working for Google is out of the question for me even though I've attempted to slip through their screening process a number of times.

The more engaged human resources are upfront in the strategy, the better they're positioned to help the company win in the market through human capital. What has to happen is that the same level of tenacity companies place on hiring the next Steve Jobs, should be placed on hiring the next set of project teams. Don't get me wrong, having the right leader in place is a start, but a leader cannot drive a company to profit if his or her team lacks skills. You have to have project athletes to compete, otherwise, all you end up doing is throwing away money on projects that are behind schedule, over budget and destined to fail.

Another key part of the human resource strategy is mentoring of newbie project managers with seasoned veterans, an apprentice program for PM's. Some companies have individuals they call Center of Excellence (COE) Leaders who are supposed to be the best at their craft. Therefore, part of the strategy is having the COE or the best corporate project managers teach the ropes to the up and coming project managers. There is nothing better than on-the-job training by someone who has years of experience. This pathway could also be promoted throughout the company as a stepping stone to higher levels of management and money!

Some professional coaches as well as managers also enjoy a gift as human resource savants. During the Dallas Cowboys Super Bowl years, Jimmy Johnson drafted and traded for athletes, looking to drive team speed, quickness and overall athleticism. He was smart enough to know that it's hard to teach athleticism (like trying to teach height). However Jimmy Johnson was confident he could teach these players the game of football. His philosophy worked and Dallas won back-to-back Super Bowls in 1993 and 1994. The Cowboys added another Super Bowl victory in 1996 under Barry Switzer. My team is Da Bears (yes, The Chicago Bears!), who should have definitely won more Super Bowls in the mid-1980s, but they were grossly out of "Shape" after their win in 1986. Team shape (focus) was lost as Da Bears rested on their laurels and didn't put in the necessary work required to win another Super Bowl.

The problem with most corporations is that hiring is usually based on skills needed for a certain function: accounting, engineering, operations or IT. Rarely is there a "How well will this individual work in a team environment?" set of interview questions to select those candidates that come with both technical and team credentials. What companies then find out is that having a surplus of brilliant scientist doesn't automatically mean that putting them on a team will lead to the next great product breakthrough - especially if egos, lack of communication, and constantly changing designs are the standard mode of operation.

Institutional Learning

Institutional learning with a feedback mechanism is getting to the root-cause of the problems as well as successes of the project. This is what everyone learns in a quality course. It's at the heart of continuous improvement, a very simple process to follow, but difficult because companies never stop to assess past projects in order to improve future ones. In fact even team members that continue onto other programs will sit idly by as teams make the same mistakes over and over. There is something very wrong about this practice. Maybe it's easier to complain about the stupidity that's interwoven in the team versus the difficult challenge of actually changing their behavior.

However, if there was a formal process for documenting project learning and updating how teams are selected, guided, motivated and staffed - the organization not the individual becomes the change agent. Better yet is if management bonuses are based on uncovering some profound knowledge from the postmortem process. If the process is well done, it makes project learning intrinsic and a way for teams to bring up the could've and should've that would've made life easier. It is a formal process for root-cause analysis; that will enhance future programs - assuming that teams truly get to the root-cause!

Corporate DNA

A lot can be said about the overall character of an organization in terms of how well they stay in project shape. A great example is any one of our amazing military branches, The Marines, Army, Air Force and Navy. It is a part of their DNA to be in the best possible shape to defend our country. Doing so is everyone's job without question. In fact, joining anyone of these branches requires this mentality from the get go. Another aspect of the military DNA is the totality of training that makes sure the whole is greater than the sum of its parts – protecting lives means that each individual must put their self aside and work together in order to face the enemy.

How does an organization create corporate DNA that drives teams to continuous improvement to succeed, knowing that there is no such thing as perfection except in its pursuit? It must start with the leader of the company to mold the DNA as he or she sees fit in order for the company to excel. Leading by example is showing everyone in the organization that they're part of the company's pathway to a success that demands consistent focus and preparation for challenges. Challenges of course posed by global competitors and constantly changing customer needs.

Just because a company was successful in the past doesn't guarantee future success. Eastman Kodak once a darling of Wall Street employed well over 130,000 people worldwide and had revenues in the neighborhood of $15B. Today they have around 25,000 people and the company has approximately $7B in revenue. Twenty years ago, Sears was not concerned with a small upstart in Arkansas called Wal-Mart. Wal-Mart is now a major player in the global economy, as Sears tries to survive in partnership with Kmart – yes Kmart.

There are those few corporations whose heritage drives everyone in the organization to stay in shape or face the exit door. One is GE who as part of their strategy purges the lower 10% of performers – translation, if you are not in project or job shape, you'll be destined for a short walk out the door. Some companies hire like-minded people through an extensive selection process such as Google, while others combine both of these attributes (highly skilled and motivated employees) as Apple has done under Steve Jobs. If you have read anything about Jobs you know that he pushes or wills his company towards leading edge products and has no problems moving obstacles to success.

Conclusion

The process is basically three reinforcing drivers that help keep teams and the organization in "Project Shape." Ingredient one, an Apple like product strategy, ingredient two, Google's ability to hire skilled people and ingredient three, The Army's institutionalization of training in team work while driving continuous improvement. "Project Shape" ends up being the result of training workers for current and future projects, constantly providing opportunities for enhancing strengths and eliminating weaknesses. It addresses all aspects of the business as well; targets in terms of new products or services, inputs needed to reach these targets and guidelines to achieve results. As well as a feedback mechanism to capture items that need improvement.

Corporations must make staying in "Project Shape", an integral part of their strategy if they want to position themselves for success. Success is defined as a more efficient and less costly development and launch process, a greater library of Intellectual Property (IP) and a wealth of talented, team oriented employees with the ability to tackle challenges. The most important part of staying in project shape is making it a part of the ongoing corporate strategy.

Back to Tiger Woods, who still goes to Bruce Harmon to work on what most believe, is the perfect golf swing. Tiger knows better – there is always room for improvement. "Project Shape" is the state teams strive to achieve that will make them successful. It's a never ending quest for perfection. Apple, Google and The Army never rest on past success, knowing that there is no such thing as perfect; so that the pursuit of perfection must continue ad nauseam.

Phenomena

No matter how much you plan for a project, how experienced your team and even the fact that you've travel this road before, problems are always part of projects – it's just part of their gene pool. The unknown, unplanned, unimagined and unbelievable disrupt projects often without warning. Project managers, at least good ones, attack these phenomena head-on, others without the stomach to face these monsters, just curl up in the fetal position.

Project success is simple, it's never about doing the mundane well or even great, it's all about how you manage problems. So much emphasis is placed on going through the motions, planning for the perfect in an imperfect world, that when something goes wrong, the project could be permanently derailed. Teams that manage problems with the precision of an emergency surgical team - triage the issues, work the critical ones, find the root-cause, fixed the problem and move to next one, these are the teams that win the project game.

Murphy is the name team's use as the cause for some unforeseen problem, so what can be better than to talk about how project managers can deal with him man-to-man or woman-to-woman. Project managers live and die by how they handle to unexpected, and beating Murphy to the punch is one way to have the advantage. Can you predict any and all things that can go wrong, of course not, but there is no excuse from being prepared.

Can you Manage Murphy?

Introduction

All projects have had that unexpected visit from none other than Murphy, as in Murphy's Law – "Anything that can go wrong will go wrong." Murphy appears as un-forecasted rain on your wedding day (can you say Alanis Morissette), hungry mosquitoes without Off on a camping trip, late software on a major program or Mom and Dad, I want you to meet my fiancée, Charles, who just got paroled from the federal penitentiary. In life as in projects, most people deal with these occasional but always difficult problems once they occur, rationalizing that there is just no way to read Murphy's mind.

Even though Murphy is defined in the context of what can go wrong, will go wrong, what does it mean in today's business world? It means missing time-to-market goals with over budget projects that allow customers to spend money on the competitor's products. It's really anything project related that gets in the way of a company making money. Time and again, projects even with the best of intentions fail to achieve goals and objectives set early on. Yet, we continue to repeat past history, can you imagine Bill Murray in Ground Hog's Day doing the same things over and over without learning what not to do from the previous days – if so, you have the typical project.

So here is the question to all project managers – "Can you manage Murphy?"

This question is at the heart of every single project manager who has dealt with problems in the past and told themselves, it will never happen again. But then a new project is handed to you and the same old problems arise, marketing is constantly adding product features, design is late to get the software completed, quality finds a major problem weeks before launch and operations is struggling to build units as a handful of key components are past due. Even with the advantage of trying to circumvent these issues in the beginning by taking into historical actuals from previous projects, many project managers do not consider Murphy in their schedules. Adding insult to injury, upper management usually forces unrealistic schedule dates arguing that current ones (realistic ones) will negatively impact targeted revenue.

What usually occurs is that the predicted end date (based on historical actuals), ends up happening anyways which is months later than management's target. However, this reality is overlooked and blamed on the common problems that the project manager tried to address in the first place. Managers never look at what it actually took to perform the project as a whole just the problems that caused the delays, whereas reality is always somewhere in the middle of these sometime extreme end points. So the vicious circle of shortchanging the project and not preparing for Murphy gets passed onto to other programs and becomes a part of the culture.

After millions or even billions of projects over mankind's existence, you would think that dealing with Murphy would be a normal part of everyday life and projects. Adding lead-time to a part in an MRP system that's always two weeks late or ordering safety stock of a key component are ways of dealing with problems, but these fail to fall into the Murphy domain. These are small potholes that could be eliminated by plain old common sense and are not Murphy worthy.

Back to the question posed earlier, can a team really be positioned to deal with Murphy or better yet, be so well prepared as to kick his backside? If you have been on teams for an extended period time in your career, there is a chance that one of those teams was able to take on Murphy and win, while others failed miserably. So what enabled some teams to succeed and others to fail?

Experience – Nothing Like It

You can't teach experience but one of the best weapons against Murphy is a seasoned group of project veterans, people who have been there and done that. Look around at successful teams and a majority probably had individuals with countless good and bad projects under their belt. What comes from past projects is knowledge that can be applied to the next. It's remembering working 24/7 to address a software timing problem and not wanting to repeat this situation on a new program. Putting together major subassemblies only to realize that tolerance build up wasn't considered and nothing fits together. Or better yet, rushing to ship a product to meet a shipping goal and getting called a month later that unit wasn't working, only to have the service engineer discover that the unit was missing a power supply. [Yes, the missing power supply story is true!]

Experienced teams don't panic at the site of Murphy knowing that they have met him or her in the past and came out on the winning side. Sometimes these boxing matches went 15 rounds, other times it was a TKO in the first round. Practice makes perfect goes for pretty much anything in life including dealing with a project roadblock that was never anticipated in the beginning. Experience provides individuals project memory analogous to muscle memory in athletics, so that they are able to meet any challenge. In fact some individuals can visualize problems in advance, as athletes such as Larry Bird, Wayne Gretzky, or Diego Maradona were able to anticipate moves on the court, rink or field to the advantage of their teams.

Once upon a time, I was part of team that thought a major new product was ready to ship after what was perceived as extensive internal software testing. The Quality team tested the product on a multitude of platforms both PC and Mac as well as numerous operating systems and the software performed as expected. Fast forward 6 weeks later with thousands of units in the market and calls into our service support center about software crashes that had customers ready to throw the product in the dumpster. The Quality team went back to the interoperability lab and could not repeat the problem.

A team of people gathered to figure out the next steps when someone mentioned a novel idea, asking customers the exact configuration of their system. After getting data through the support center and determining that all met the minimum system requirements, a discovery was made; all problem systems had Microsoft Office (D'oh!) which caused a conflict with our USB driver. You guessed right, we never tested compatibility of our application with other software packages, since the interoperability lab was only loaded with our software. It took no time at all to experience what customers were dealing with, major crashes that caused the user to periodically reboot the computer. Add insult to injury, both the software and firmware needed to be updated – Murphy kicked our butt.

After this experience we sent a questionnaire to customers to understand the types of software they typically had on their computers, added these to the interoperability lab units, increased the number of external beta test sites and more extensively tested future software versions on current and new products. In addition, we developed a check sheet for service support to complete when customers experienced software problems. Instead of getting flustered by customer problems we looked at it as a way to make the product better. All of these actions based on experience went a long way towards improving the quality of our software on the next program.

Bring it on, experience provides confidence in handling, even daring Murphy. Quarterbacks like Joe Montana, Dan Marino and John Elway actually enjoyed taking on Murphy and, in fact, thrived on adverse circumstances with confidence that they would win the game in the final minutes. The same goes for teams that have years of immeasurable proficiency handling problems. They don't get flustered by problems, in fact the imperfect nature of projects is expected and fits into what just needs to be addressed as part of the project.

Time is your friend

There never seems to be enough time to handle all the tasks required to successfully manage a project, let alone reserve time for those un-expected problems. However, project managers have ways to make time their friend by creative planning. And the best way to stay ahead of Murphy is by spending adequate time in the beginning of the project planning for what is needed for victory. Paying later versus paying now will only cause problems downstream in a project. Program managers must not embark on a project unless they have prepared in advance for what is expected in the end. The best way to make time your friend is plan, plan, and plan, just like the key to a successful restaurant is location, location, and location.

Once the outcome of a project is truly understood, the project manager can look for opportunities to reserve time to deal with Murphy, think of this as Murphy Slack. Experience shows that you don't know what you don't know until you know what you didn't know. Once you know what you didn't know, you can take action to correct the situation. And the best way to do so is to get started early before it becomes increasingly difficult to redirect the team onto an alternate journey. Case in point, once long before GPS units where available, my father listened to a relative about how to get to our final destination only to travel over a 100 miles in the wrong direction. Let's just say that on the rest of the trip, the tension amongst the 3 adults and 5 children could've been cut with a knife and a very sharp one at that. With today's GPS system, you would get the obnoxious "recalculating" audio which really means you are going the wrong way stupid. Too bad teams don't have electronics to warn them of Murphy's impending appearance.

As the Beatle's song goes, I get by with a little help from my friends, which is a great mantra for project managers. Friends in the form of suppliers, customers, international regions, anyone who touches the project team who can provide additional capacity and more importantly extra task hours. Project managers are never truly given the appropriate number of resources or timeframe to mitigate Murphy, but good ones take advantage of key suppliers, customers and regional staff to provide a buffer for the unexpected. With more resources at their disposal, problems can be discovered earlier as tasks are done in parallel and before major changes that could derail the program occur. By intelligently delegating tasks, the project manager can better prepare for the unforeseen.

I recently needed to have four new circuit boards designed, built and tested on a very short turn. In partnering with a local supplier, we took advantage of their ability to review boards for manufacturability, testability and panelization of bare circuit boards for cost. In addition we utilized their purchasing department to order parts, stock parts and find substitutes when ones that were on our bill of material had extensive lead-times. In essence, our small team of 10 people added over 20 more members from our supplier and did so in a much less expensive manner than temporarily hiring more staff. We created more task time by delegating items that the supplier did well, while freeing time for our engineers to design the boards. Without this partnership, what took 6 months could have easily extended to a year and killed our chances to win a major contract.

Camaraderie

Even though obtaining the nirvana of team camaraderie is a rare achievement, the collective skills and experiences of a team always pose a challenge to Murphy. Team dynamics are critical because instead of one or two individuals setting out to fight Murphy in hand-to-hand combat, the force of the team can be used to kill this foe. As much as I believe Superman is the baddest superhero in the world and can take on any nemesis, it certainly helps when other Justice League heroes help him battle intergalactic villains. The Justice League brings other heroes with complementary skills that once melded with Superman's gives them the advantage to win over evil.

It's too bad there's no magic formula for team selection to guarantee camaraderie, but when it happens, it's a beautiful thing. Only a handful of times in my career have I enjoyed what could've been called true camaraderie and during these experiences Murphy never had a chance. On one project an image quality problem surfaced just months before launch, causing an email to management about a potential delay in meeting the product launch date. The team didn't even flinch, instead leaving the meeting together to work on the problem. With the precision of a surgical team, they discovered an out of specification part in the imaging subsystem which caused a timing error in the software. Within hours the problem was solved and testing resumed. This effort was followed up with a visit to the supplier to understand the root cause of the part being out of specification so that the issue could be permanently fixed.

On the flip side of camaraderie is the team with one member infected with H1N1, the virus that will destroy team unity and success with limited chance for recovery. It doesn't matter if every other person is a "team player"; one bad member will ruin the team's ability to manage problems, especially of the Murphy variety. Those television commercials showing the coughing H1N1 individual spreading the virus to others in that nice shade of green goes the same for teams. Murphy knows where there are weaknesses and that the true strength of the team is directly correlated to the weakest link. H1N1 infected members pose a great opening for Murphy to attack and minimize the effectiveness of the team.

Frankly having experience with teams ripe with H1N1 members, the best thing to do is isolate them so they don't infect the rest of the team. Or better yet, take them off the team forever. Of course you can take the time for the H1N1 member to get well; however, in the business world there is no vaccine for these naysayers and many times the team can't afford the time and money it will cost to make them better. The real key - do blood work on everyone prior to adding them to the team, if they fail the test, put them in quarantine immediately.

Don't Open the Door Stupid!

In every scary movie there is always an actress or actor who opens the closet door even though they know that some killer is lurking behind the door, only to be killed in some utterly gory and violent manner. As an engineer with some logic, I cannot believe that writers and audiences don't stop this constant act of stupidity, unless the real thrill is seeing how these B actors will come to their demise. So some sage advice for project teams, don't open the damn closet door and make the same mistake twice or three times that allowed Murphy to win.

Now this advice might sound like a no brainer, but with the level of technology and time-to-market pressures, in teams zest to beat the competitions, they have memory lapses. For example, for anyone that has had the pleasure or pain of trying to get a unit to pass Electro Magnetic Interference (EMI) regulations for a product, you know it is part science, part art, where the art portion can give Murphy a window to enter. I've seen months go by with very smart engineers wrestling with one of two frequencies that fail to meet either a US or international EMI threshold. Or software bugs that somehow manage to get through all of the testing only to be discovered by the end customer. Worst yet, a drug that passes FDA regulations that caused unwarranted problems in patients, even death. Anyone who watches television will see law firms advertising about known problems with drugs in order to cash in with class action suits. So why did these doors get opened with the killer ready to strike?

Part of not opening the closet is planning an alternative way out of the house and the best time to plan for Murphy is upfront. Take software testing, Windows Vista had numerous problems; however, Windows 7 appears to be better in terms of reliability and features. So what did Microsoft do differently? My guess is that Windows 7 is really a version of Vista with all known problems fixed and some added features to mask Microsoft's previous failed attempt. After all, the more time a product is out in the market, the more time customers have to discover problems, shortcomings and other features they would like. Windows 7 is a combination of all of these years after the initial launch of Vista. It appears that Microsoft finally found a better way out of the house and didn't get killed by the Boogie Man – irate customers.

Communications or Lack Thereof

As a youth, I tried to take advantage of the tennis match between my parents when asking for permission to do something they normally would never agree too. This came about through years of trial and error in determining those situations where they did not communicate about their respective decisions. "Dad, can I go Mike Nixon's house for a party tonight?" Answer "His parents are out of town this weekend, which means no way!" "Mom, dad says he doesn't see a problem with me going to Mike

Nixon's party with our friends." Answer, "I guess if your father's okay with you attending the party, its fine me." There you have it, a little misdirection, knowing (or hoping) that the lack of communication between my parents would allow me to attend a party at Mike's house while his parents were out of town.

What does this have to do with Murphy? Many times the only reason he succeeds is because team members fail to keep each other informed of their actions or worst yet, problems. Or two team members assume someone else is performing a critical task, for example ordering a long lead electrical component only to find out 8 weeks later that no one ordered the part and now the program will be delayed by 8 weeks!

There is more and more technology around that deals with Murphy, for example GPS units and or Google maps that can pin-point constructions zones to prevent the "Are we there yet?" rant from bottled up children on a cross country vacation. I don't know about you, but bypassing a construction zone during a hot and steamy summer day with 3 children in tow works for me. Mercedes even has a vehicle to prevent sleeping drivers from rear-ending cars in front of them or crossing over yellow lines – leading to saved lives. So clearly there are some inventions that go a long way towards preventing disasters, many that are caused by human error.

Conclusion

So can you manage Murphy? Well the answer is yes, there is a way to either prevent Murphy from entering your project or kicking his backside if and when he arrives. The best way to do so is to never ever believe there is such a thing as a perfect or smooth running project or one that will not make the same mistakes as in the past. Expecting the unexpected is what good project managers do – period. In fact predicting what will happen on projects is like predicting the weather, even with all of the computer models and high technology satellite imaging the forecast is just that, an intelligent guess.

Teams that somehow believe all is well give Murphy the opportunity to wait for just the right time to push the critical path to the edge of no return. However teams with experience, camaraderie, and the support from suppliers can circle the wagons and take Murphy's best attack. In the end, the ability to be calm, focused and tenacious to defend the critical path will lead to overcoming anything that Murphy has in his arsenal. The best way to handle Murphy is to know that he has showed up on every project since the beginning of time – can you say Adam and Eve!

Everything is Interwoven

As you read through the various articles, it should come as no surprise that people, parts, processes and phenomena are all interwoven and trying to separate one from the other is an impossible task. In fact, people are probably the cause of problems with parts and phenomena, and responsible for developing the processes that we all love to hate. Success for any program manager depends on all of these ingredients working in unison, a feat that is rare if at all possible.

The following articles were hard to place under one of the The PMPs, so they have been put here as a way to admit that fine lines between people, parts, processes and phenomena can be difficult to define. This is exactly why being a project manager is hard, which of the four Ps needs to be pushed, pulled, polled, prodded to move the schedule forward. How do they interact, will changing one negatively impact the other, what happens if you change the wrong one, etc. The best way for a project manager to be successful is to never take his or her eyes off the project. Project management is analogous to driving down a snowing Buffalo highway going 65 mph; it takes but a second of lost focus to crash.

There's no Time for Eggnog

Introduction

Two weeks ago, a very good friend of mine and a really smart person mentioned his frustration with some of our peers. How during the course of a critical project, that the project manager's focus was more on when he could put up holiday decorations and partake in some hearty Eggnog, than meeting the deadline of the project. Let's just say that his words were much more colorful than what I've written, however, his point is well taken. Bottom-line, shareholders don't really care about your personal life if it gets in the way of maximizing their wealth. As a project manager, your life is not your life, it's the shareholder's. And to assume otherwise will just make you average at best in your profession.

As a member of this same team and an experienced project manager, my focus was trying to do the best job possible to help our company. Having been downsized twice, I realize that sometimes you have to put your personal life on hold so that you can maintain your job and pay the bills. This was one of those times. In parallel, others in the organization used this opportunity to pad their wallets with overtime (OT) or decided that losing Paid Time Off (PTO) was not an option and let others take on the additional burden of completing this poorly timed project.

Don't get me wrong, personal time in the form of chaperoning a school trip or taking the children to the home of "The Mouse" are needed for your sanity, but too often, employees forget that the money we are spending is not ours or our managers. Shareholders provide hard earned dollars to corporations with a clear objective, making them more money! If making them more money comes at the price of some personal time, so be it. This is even truer for project managers.

Project managers are constantly in no win situations and are the first head to be placed in the guillotine. Therefore they are going to have to make difficult choices that impact workers personal lives. How and when to make these calls and have workers stay late or over a weekend without extra pay is akin to the boy who cried wolf, you can only do it so many times before your team eats you. Certainly no one wants to spend the majority of Eggnog season working, however, if it means that you can pay the credit card bills after the holidays end, sign me up.

How do project managers direct teams and make the calls on when it is appropriate to drink Eggnog and decorate the house versus staying late at work to complete a major proposal? The first step is making sure that you as the project leader set a very high standard for what is expected from the team.

Lead by Example

"With great power comes great responsibility" (Stan Lee of Spiderman fame), is the mantra all project managers must uphold. And with such high esteem, project managers must lead by example. There can be no downtime, no complaints about marketing's lack of intelligence or management's overzealous commitments; his or her focus must be on the success of the project even at a personal cost. The project manager's job must be 24/7/365. There is no rest for the project manager, they must be the role model and can't take time off like Michael Jordan did during the NBA Finals to gamble.

Once upon a time, one of my upper level managers decided to go out to lunch at a club house with a beautiful view of the golf course. To his surprise, the project manager of a late and over budget project was just outside the restaurant window taking golf lessons from the local club pro – this is definitely not leading by example. Let's just say the project manager never lived this event down from his peers and never again managed a program. If you're going to improve your golf game, do it over the weekend, not during work hours in full view of the person paying your salary, benefits, and yes, golf lessons.

It's amazing how simply leading by example can set the expectations for team members about when it's time to hit the spreadsheets versus spirits. There should be no one on the team that works harder than the project manager. If this is not the situation, the damn will open and team members will start an uprising about why they have to do this or that, while the project manager is not accountable for exhibiting the same behavior. The project manager cannot go on vacation in the middle of the major crisis or critical part of the project and expect the team to understand this action. If you need an example, look no further than New Jersey Governor Chris Christies' extended Disney vacation, while the state was digging itself out of a massive snow storm.

As the old adage goes, "Do onto others as you would have them do onto you" works for project managers. Model the behavior you want in your team and this will be reciprocated tenfold. Taking another approach will send members spiraling into the abyss of complaining, slacking off and moving the project towards a cliff. It's hard to always be on top as the project manager, but this is the first step in keeping people's mind off leaving early to shop for tree ornaments, while the competition is working hard to eat your breakfast, lunch and dinner.

Human nature is to observe and imitate others, especially those in authority. People want to fit in even well past those awkward high school years and this continues into adulthood and the corporate world.

Family Backing Helps

In order to be on your game, working long hours, constantly checking the Blackberry, listening to phone messages and barely watching your son or daughter at the weekend soccer tournament takes the backing of your family. The saying behind every great man is a better woman or vice versa, holds true for project managers. Their family has to understand that sometimes dad or mom is going to be occupied with work and might not be around some nights to read a bedtime story. But once there is an opening, nobody can party like a stressed out project manager – or at least that's the rumor I'm starting.

I was part of the fore mentioned project and was not thinking of Eggnog, but doing anything I could to keep my job. In fact, one evening after a nice run and dinner, it was time to visit the in-laws for some dessert. Minutes before heading to the car, the phone rang and my wife answered – "It's for you, someone from work." With almost 28 years of marriage under our belts, we have this way of knowing what was coming and when I said, "I need to go back to work," my wife understood without questioning the sanity of my co-workers. No joke was being played; I hadn't concocted the call in order to get out of visiting my in-laws. I really like my in-laws and my mother-in-laws desserts are to die for as is her lasagna.

With the full backing from my family (okay my wife), it was off to work for a late evening with my peers to fine tune some critical work. The sole focus of the evening was on work, which helped tremendously in doing the best I could do. Family backing comes in different flavors, paying bills, running errands, cutting the grass, taking out the garbage, going grocery shopping, recording your favorite show, preparing a warm meal for a late night and making the limited time during projects special. There is no way a project manager can excel without help in the form of family support and backing.

Intuition

Great project managers are like great coaches; they have this sixth sense about how to maneuver players within the game. Unless you were cloned, most of us have experienced mom's intuition about when we were feeling under the weather, hiding a bad grade, lying about how Suzy's parents will be home for the weekend sleepover or not owning up to that small fender bender. They just know how to read us and get to the root of what's going on in our life. Maybe this training was started as an infant from their moms, part of the female DNA, some magic passed down from the ages or a combination of biology, education and life experiences? It really doesn't matter how this skill was obtained, what's great is how Mom's use it to help out their children.

Intuition in project managers allows them to know when it is time to take the team out for beers and chicken wings or hold them hostage over a weekend to get a critical item complete. Each of these is a polar opposite way to keep the team motivated on the ultimate goal. The truly great project managers understand the balancing act of pushing the team forward to the benefit of shareholders and providing opportunities for them to refresh and recharge. Is there time for Eggnog, of course, but the time must be when it doesn't hurt the pocketbooks of the shareholders. One of my mentors once told me that what I do at work should be done as if one of our major stock owners was observing my behavior - besides when nature calls.

The best way for project managers to utilize intuition takes another step out of the mom play book; know your team as if they're your children. Everyone is different and what works for one person could have the total opposite effect on another. Each person has a totally different work engine, some can drive long hauls without taking a break and others need to stop more frequently to rest, unload and reload with food/drink.

Know Your Children

Mom's not only have an unbelievable sense of intuition in regards to their children, but what truly motivates them and pushes them into the off versus on position. Project managers must understand each and every one of their children - the members of the project team. As well as key suppliers, managers, and customers and any other inputs to the team or receivers of the final output. When a project manager gets a handle on his or her children, they can start to understand when to push, when to back off the throttle and when to go for beers.

Most project managers don't have time or the patience to create multiple approaches based on the uniqueness of team members. They interact in only one way with them; this is most evident in the military. In the military your motivation is driven by how well the unit works together and the leader of the troop is not concerned about individuals but the collective group. The only time anyone drinks Eggnog in the military is when the leader commands the troop to consume this holiday beverage.

In business, understanding your children is analogous to a NASCAR driver knowing his car. By having this information you can get the most out of the team. Let's be honest, there are some project managers that want to sing Kumbaya with the team, but the sole purpose of knowing your team is to push them to the limit and then provide downtime before starting again – just like what a NASCAR driver does to his car. Truly recognizing the different motivators of people (their likes and dislikes), provides the project manager with the controls he can maneuver to cross the finish line first.

Here's the dilemma with this concept, can the project manager afford the time needed to get acquainted with his team and still meet his or her other duties. Is the time spent getting to know the project team more than offset by the benefit of being able to push them at critical times?

Clearly some of this getting to know team members is taken care of by the functional hats/skills people bring to the project, but knowing them goes much deeper. Project managers can also take previous experience with these individuals on past programs in understanding their limits. Bottom-line, anything that the project manager can do to better comprehend the limits his or her team can be pushed too in the beginning of the program will pay enormous benefits in the long run.

Trade the Drinkers

Let's assume that you're the perfect project manager and never partake of Eggnog at the holidays but just can't get everyone else on board. Well the success of the project manager is paced by the weakest link and if team members are spending critical project time planning for an Eggnog gathering, it might be time to lose them. In the end, it doesn't really matter if you are or think you are the pinnacle of project management. You will get blamed for late deliveries no matter what you do in order to change the behavior of the slackers.

It's certainly reasonable in today's world that not all people have the same drive as you in the never ending quest to achieve the impossible; a project on-time and under-budget. However, if the team is not performing up to par, it's time to replace them for others that will only drink Eggnog after the project has ended. Or as in the military, when the project manager says it's time to drink.

Of course there are times when some of the drinkers have such valuable skills that some delays are more than compensated by the quality of the work. The best way to get around this is to plan extra time in the beginning before committing a firm project end date to management. However, for those that provided durations and due dates but find it hard to meet them given a focus on outside activities, it's time to move forward without them. Not taking this drastic step will end up being your problem when faced with upper management grilling you on a late and over-budget project.

Conclusion

Managing projects is a pain! You and only you have to somehow take a group of unique individuals and get them all to perform as one even if it means loss of personal time. Unlike driving a NASCAR vehicle to its limits, attempting to do the same with humans could get you in a heap of trouble. You can certainly lead by example, get to know each member, use your intuition and trade away the drinkers, however, this will never guarantee success. The other option is to know that to push people to a greater good takes someone who really doesn't care about winning a popularity contest. Even with the above tools in your arsenal, project managers have to push their teams with the expectation that they will receive few if any invitations for holiday Eggnog.

For me it's very easy to get everyone on the same page, just ask them if this was your company and money, how you would expect others to spend it. This is what I believe is my role as the project manager, making more money for the company - plain and simple. My philosophy is based on knowing that it's certainly easier to enjoy life outside of work when you are gainfully employed, than putting my company in a position to lose out to a competitor. Others place more importance on their life outside of work and that is certainly a fine approach. However, when these two opposing life views collide, there can be only one winner – the project manager. Life comes with no guarantees and even if the project manager does everything in their power to keep their job, nothing is a hard and fast.

So What?

All project managers see the world through experience, figuring that training without real hand-to-hand combat doesn't make them any better. However, there is something that can be said for understanding that we all need support, even project managers. What better support system than other people with similar experiences that can provide some guidance before, during or after the battle.

Hot Chili and Cold Beer for the Project Manager's Soul is just a way of sharing war stories with lessons learned. Will this solve all of your problems as a project manager? I'd be lying if I said yes. But it should at least provide reassurance that in similar circumstances, your approach was good – maybe you just didn't have the right mix of people, parts, processes and the phenomena kicked your butt.

In the end, project managers need little in between projects, yes some hot chili and cold beer, and then we're right back into the fight. So what? Continue to push forward as the world depends on project managers to fuel the economy with new products and services. Pundits say small businesses are the engines of the economy – I disagree – project managers are the real engines of the economy.

Resources

Friends from C Building

Web Sites

Eastman Kodak Company
http://www.kodak.com

Books / Articles

Crawford, Merle, Di Benedetto, Anthony, *New Products Management*, McGraw-Hill, 2006

Katzenbach, Jon R. and Smith, Douglas K.: *The Wisdom of Teams*, Reprint Edition, Collins, 2003.

Leavitt, Harold J. and Lipman-Blumen, Jean: Hot Groups, Harvard Business Review, Reprint Number 95405, July-August 1995.

About the Author

Donald Pillittere is an accomplished manager with proven success in developing and implementing long-term strategic vision to achieve profitability. His leadership has enhanced profit performance in a variety of disciplines both by increasing sales and significantly reducing expenses, without sacrificing quality or customer satisfaction. These results have frequently been achieved while simultaneously beating time-to-market deadlines. This results-driven professional's 30-year career has encompassed digital imaging, medical imaging and health sciences; and process controls for both light manufacturing and heavy industrial applications.

Currently a Subcontract Program Manager at ITT, Mr. Pillittere manages major procurements for highly advanced subsystems. Prior to working at ITT, he was Director of Product Engineering and Program Management at Transonic Systems Inc., where he managed the development and introduction of numerous medical flow measurement products for both clinical and research applications. Previously as Worldwide Product Manager for the Eastman Kodak Company, he launched numerous award-winning scanners that significantly outperformed sales projections and exceeded profitability goals. He also has a successful track record in the international arena, with experience in Asia, Europe, Latin America, and Japan, as well as North America. A sought-after expert, Mr. Pillittere was frequently quoted in corporate press releases, as well as various periodicals and professional journals.

Sharing his wisdom and real-world experience as Adjunct Professor at his alma mater, Mr. Pillittere has taught graduate courses in Operations Management, Supply Chain Management, Manufacturing Strategy &

Tactics, Managing Manufacturing Resources and Engineering Economics at the Rochester Institute of Technology since 1999. He successfully combines his industrial experiences with the theory and principles of operations management to offer students practical, proven tools and approaches that can be applied in a wide variety of business settings. A gifted public speaker, Mr. Pillittere creates a dynamic, challenging, and fun learning environment that he adapts to the specific needs of each class.

He earned a Bachelor of Science degree at the State University of New York at Buffalo in Electrical Engineering; and holds a Master of Business Administration degree from RIT's Executive MBA Program, from which he graduated With Honors, as a member of Phi Kappa Phi and Beta Gamma Sigma Honor Societies.

Clearly a highly motivated individual in his personal life as well, Mr. Pillittere is an accomplished marathon runner, having competed in events in New York City and Chicago. Happily married for 29 years, he and his wife are the proud parents of three children, with whom he resides in Spencerport, New York.

Reviews for
Are We There Yet, Diary of a Project Manager
(2nd edition coming soon)

Amazon.com

5 out of 5 stars – All too familiar episodes
By Lino Nobrega, March 28, 2010

"I found myself smirking while reading this book, it takes you on a ride through Corporate events, that are probably common in most work places. Well detailed, and I recommend that you learn from Don, he knows what he's talking about. Buy the book for your entire team, not only will they appreciate it, they will get great insight and hopefully avoid some of the common mistakes."

5 out of 5 stars – A book for project managers and team members stuck working in the real world
By Gene R., April 10, 2011

"Having read numerous books on project management methodology, technique and practice I found Don's book to be practical and realistic in regard to what ACTUALLY happens in the REAL WORLD of project management. Don't get me wrong, as a certified PMP I think the PMBOK (Project Management Book of Knowledge) is a great resource describing how projects should go. Don simplifies the process into what he calls the four "P" of project management People, Parts, Processes and Phenomena. His real world stories will send the message home and will even make you laugh, especially if you have spent much time in corporate America. It is an easy read with great content; I recommend the book to anyone working in project management or as a contributor on project teams."

PM Forum
Published in *PM World Today* – August 2010 (Vol XII, Issue VIII)
PM World Today is a free monthly eJournal - Subscriptions available at http://www.pmworldtoday.net

Excerpts from review

Introduction to the Book

This book focuses on all elements of the project management lifecycle, utilizing real-life examples in a diary format that is easy to relate to and understand.

The project management 4 P's (Process, People, Parts, and Phenomena) enables the reader to become more knowledgeable by handling obstacles (internal and external), thereby providing a foundation to be better equipped to succeed. Being that each project is unique in nature, this book focuses on important aspects such as working culture, internal processes, and the team environment. Ultimately, project characteristics that is consistent among all projects.

Overview of Book's Structure

The diary format symbolizes a day-to-day chronicle of a project team developing a series of new products. This stimulating read has short chapters, providing an engaging look at the trials and tribulations of project management; accompanied by many 'aha ha' moments in this vastly dynamic profession. Each chapter is structured where the project obstacle is stated upfront. The chapter discussion highlights the current scenario the team is facing, and the chapter ends with what the team could have done better to mitigate the difficulty.

Highlights: What I liked!

The flow of the book is a story in its entirety, based on real-life experiences and is very easy to comprehend. The comical highlights add humor to stressful situations.

"One needs to see the light at the end of the tunnel...and sometimes it's an oncoming train!" Additionally, the 'character list' was well thought out, creatively written, and a 'nice to have' reference. Furthermore, the final chapters tie the entire book together, and summarize the four 4's of project management (Process, People, Parts, and Phenomena), relating examples to applicable chapters.

CPSIA information can be obtained
at www.ICGtesting.com
Printed in the USA
EDOW021246010413
1065ED